Jet Reconstruction and Spectroscopy at Hadron Colliders 2011

Journal of Physics: Conference Series Volume 323

Pisa, Italy
18-19 April 2011

ISBN: 978-1-61839-293-0
ISSN: 1742-6588

Printed from e-media with permission by:

Curran Associates, Inc.
57 Morehouse Lane
Red Hook, NY 12571

Some format issues inherent in the e-media version may also appear in this print version.

Printed by Curran Associates, Inc. (2011)

For permission requests, please contact the Institute of Physics
at the address below.

Institute of Physics
Dirac House, Temple Back
Bristol BS1 6BE UK

Phone: 44 1 17 929 7481
Fax: 44 1 17 920 0979

techtracking@iop.org

Additional copies of this publication are available from:

Curran Associates, Inc.
57 Morehouse Lane
Red Hook, NY 12571 USA
Phone: 845-758-0400
Fax: 845-758-2634
Email: curran@proceedings.com
Web: www.proceedings.com

Table of contents

Journal of Physics: Conference Series

Volume 323

2011

Jet Reconstruction and Spectroscopy at Hadron Colliders
18–19 April 2011, Pisa

Preface

Papers

INTRODUCTION

Dear colleagues and friends,

Major new particle discoveries were made in the past by exploring the mass spectrum of lepton pairs. These searches still have great potential. However, new particle searches are now being extended to masses larger than the W, Z mass. More and more decay channels open up and the branching ratios into lepton pairs are reduced. Also, physics may dictate that states with heavy bosons and quarks become dominant. Examples are the decay of top quarks, and the expected final states of the standard model Higgs boson. Supersymmetry in any of its wide spectrum of models predicts intrigued final states where jets are major observables. To reconstruct masses and to study the dynamics of these states one must exploit the energy-momentum four-vectors of jets.

Past experiments at the CERN SPS collider, at HERA, at LEP and now at the Tevatron collider and at LHC, have studied how best to reconstruct hadron jets. However, originally the role of jets in searching for new physics was primarily to sense new parton contact interactions by means of increased large pt tails in inclusive jet spectra, or studying jet events with large missing Et, or measuring branching ratios into jets of different flavour. These studies did not require as accurate a measure of jet four-momenta as needed in new particle searches in multi-jets final states.

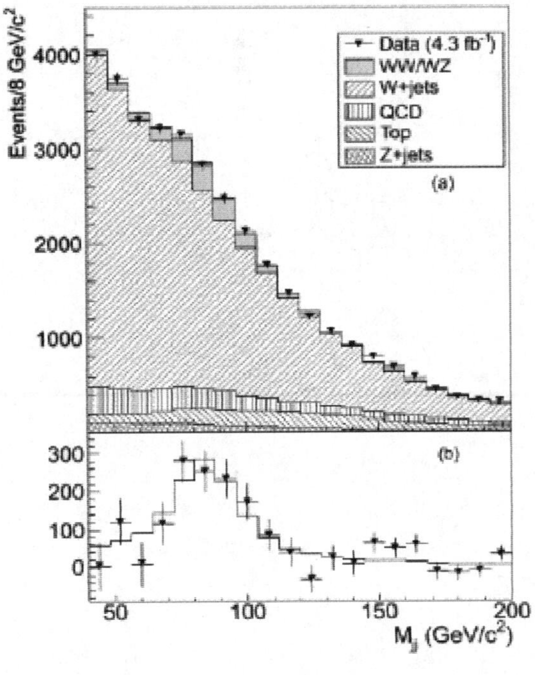

Figure 1. W, Z associated production in CDF events with large Et, miss and 2 jets.

Consider for example (figure 1) the mass spectrum of dijets in events with large missing Et recently measured by CDF [1]. Trigger and analysis cuts were chosen so as to favour production of heavy boson pairs, with decay of one Z boson into neutrinos tagging the event and another W or Z boson decaying into jets. Associated production of boson pairs is observed, but the dijet mass resolution does not allow the separation of W from Z. A broad agreement of the overall observed rate with expectation is found, but a comparative study of the two processes cannot be made.

The feeling of uneasiness becomes much worse by observing (figure 2) the invariant mass distribution of jet pairs produced in association with a W boson at dijet masses larger than the W, Z, recently studied by CDF [2]. Evidence is shown for a wide bump around 145 GeV/c2 which cannot be explained by known processes. Assuming that it is new physics, we do not know whether a single new state or a superposition of several processes is being observed. In an attempt to strengthen the significance of the result one would like to split this spectrum into jet flavour channels. Since the expected background is primarily composed by gluon jets, it ought to be different in different channels. The new process might stand up more prominently in some flavour-specific channel. However, at the present time this study can only be performed accurately for beauty-flavoured jets. Accurate flavour tagging for all jet flavours to include, in particular, gluon jets would be needed. Extending jet flavour studies much beyond the accuracy presently available is mandatory.

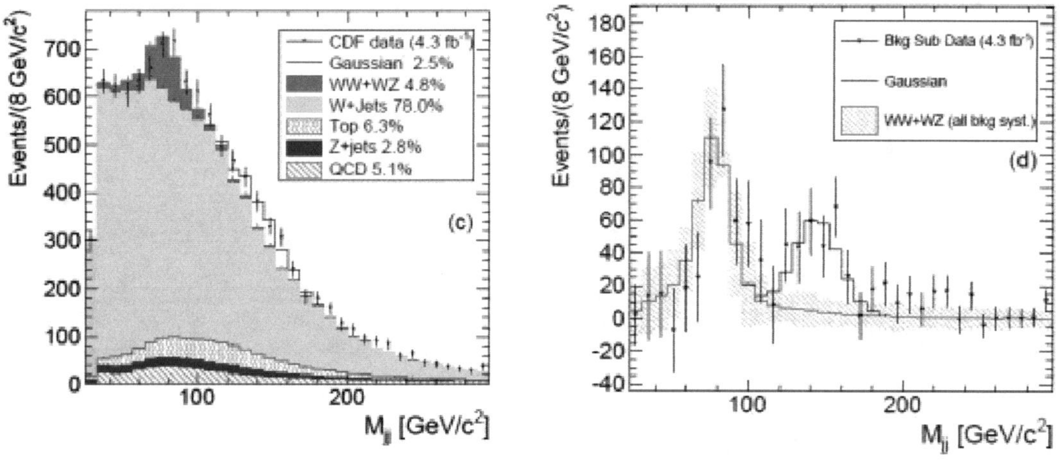

Figure 2. Mass spectrum of dijets in W production events.

The role of theory is extremely important for ensuring success to a renewed jet spectroscopy effort. Part of the theory effort would be in phenomenology. Tagging jet flavour requires an understanding of flavour-dependent jet fragmentation. This goal can be pursued by detailed measurements of jets of known flavour combined with their accurate phenomenological description. A probability tag for each flavour choice could become a standard 'observable' to be quoted as a jet parameter.

Precise perturbative calculations can be made of the kinematical structure of the final state of multi-jets of heavy particle events. This is true for W, Z and top quark as well as for new exotic particles. Theoretical input will be vital in understanding the physics of such events. Such an input might be necessary for discovering new particles.

Jet spectroscopy, as difficult as it might be, is a vital field of research. We should not give up, not at all. This is why we started this series of workshops. We would like on one hand to document in a comprehensive frame the progress being made in calibrating the jet energy and improving the jet energy resolution, and at the same time to show that the effort is being rewarded with significant scientific findings. Since this can be achieved only with a joint effort of experimentalists and theorists, theory contributions are, and will be, solicited and most welcome. I trust that our work will be fruitful to the entire HEP community.

Giorgio Bellettini
Physics Department and INFN, Pisa

References

[1] Aaltonen T et al 2009 The CDF Collaboration *Phys. Rev. Lett.* **103** 091803
[2] Aaltonen T et al 2011 The CDF Collaboration *Phys. Rev. Lett.* **106** 171801

Effects of Calorimeter Peculiarities on the Jet Energy Scale

Richard Wigmans
Department of Physics
Texas Tech University
Lubbock, TX 79409-1051, U.S.A.
E-mail: `wigmans@ttu.edu`

Abstract. It is well known that non-compensating calorimeters, such as the ones used in the general purpose experiments at the Tevatron and the LHC, offer particular problems for correctly reconstructing the energy deposited by single hadrons and jets. In this talk, I will review these problems, discuss the remedies usually employed, and possible alternatives. I also briefly review the options for hadron calorimetry at a future Lepton Collider, where jet spectroscopy is expected to become a major experimental tool.

1. Introduction

The question of the jet energy scale (JES) in modern particle physics experiments is crucially linked to the interpretation of the signals provided by the calorimeter system. The fact that a special workshop is organized to discuss this issue indicates that this interpretation is far from trivial indeed, especially if the experiments use non-compensating calorimeters, which is true for all general-purpose experiments at the Tevatron and the Large Hadron Collider. The non-triviality mainly concerns the calibration of longitudinally segmented calorimeter systems. In this talk, I will first describe some general consequences of non-compensation that affect the JES. Most of my time will be spend on describing different methods that are used to intercalibrate the sections of longitudinally segmented calorimeter systems, and their (adverse) consequences. At the end, I will briefly discuss new developments that may affect the future of hadron calorimetry in particle physics experiments.

2. General aspects of hadron calorimetry relevant for the jet energy scale

The absorption of a high-energy hadron in dense matter is characterized by two components, which we will call the electromagnetic (em) and non-electromagnetic shower component. The em component is initiated by π^0s and other electromagnetically decaying hadrons, while the non-em component encompasses everything else that occurs in the absorption process. The latter component is dominated by *nuclear* processes: release of protons, neutrons and nucleon aggregates from the nuclear environment in which they are bound. The nuclear binding energy that has to be provided to make these processes possible does itself *not* contribute to the measurable calorimeter signals. It is usually referred to as *invisible* energy, and may comprise on average up to 40% of the energy carried by the non-em shower component. Mainly as a result of this invisible energy, the calorimeter response, *i.e.,* the average signal per unit deposited energy, to the non-em shower component is typically considerably smaller than the em response. Because of the large variety in possible nuclear reactions (and associated binding energy

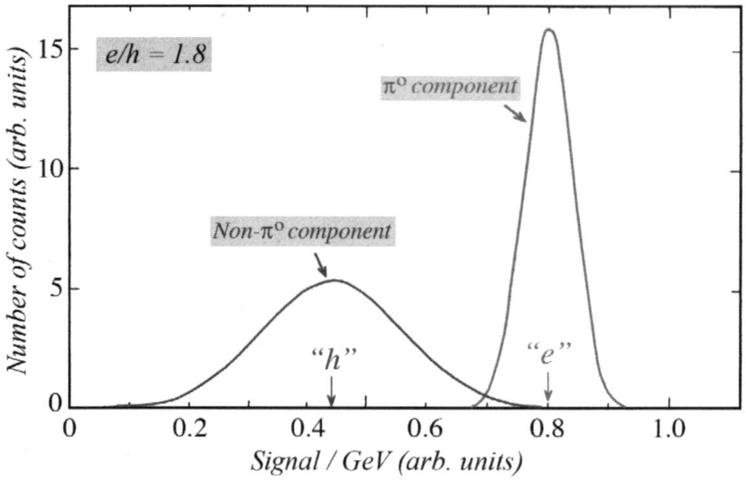

Figure 1. Schematic representation of the response functions of a non-compensating calorimeter to the em and non-em components of hadronic showers. The ratio of the mean values of these distributions is the e/h value of this calorimeter.

losses), the width of the non-em response function is also larger than that of the em one, where invisible energy plays no role. Both effects are schematically illustrated in Figure 1. The ratio of the mean values of these distributions, *i.e.,* the ratio of the em and non-em responses, is known as the e/h value of the calorimeter. In this example, $e/h = 1.8$, which means that on average 1 - 1/1.8 = 44% of the non-em energy does not contribute to the calorimeter signals. Calorimeters with $e/h \neq 1$ are usually called *non-compensating*.

A shower induced by a high-energy pion has both an em and a non-em component. The response function of the calorimeter for such pions thus centers around a mean value in between those for the em (e) and non-em (h) components, at a value determined by the average energy sharing between these components at that energy ($\langle f_{\mathrm{em}} \rangle$). As can be seen in Figures 2 and 3, the characteristics of f_{em} and its fluctuations have direct and defining consequences for the hadron calorimeter performance. These characteristics, some of which are illustrated in Figure 2, include:

- The fact that $\langle f_{\mathrm{em}} \rangle$ increases with energy (Figure 2a). As a result, non-compensating calorimeters are intrinsically *non-linear* for hadron detection (Figure 3a).
- Event-to-event fluctuations in f_{em} are non-Poissonian (Figure 2b). The hadronic response function (line shape) mimics these fluctuations as well (Figure 3b).
- As a result of the non-Poissonian nature of the f_{em} fluctuations, their rms width does not scale with $E^{-1/2}$ (Figure 2c). The same holds for the hadronic energy resolution (Figure 3c).
- The fluctuations in f_{em} depend also on the type of hadron that initiates the shower. For example, in showers induced by high-energy protons, $\langle f_{\mathrm{em}} \rangle$ is smaller than for pion showers of the same energy. Also, the event-to-event fluctuations are smaller and more symmetric in the proton case. As a result, the calorimeter response to protons is smaller, the resolution better and the line shape more symmetric than for pions [2].
- The production of em shower components is *not* limited to the early part of the shower, but may take place over the entire absorbing volume.

The latter effect, which is illustrated in Figure 4, greatly complicates the intercalibration of the different segments of a longitudinally segmented calorimeter system. Since em shower components are sampled more efficiently than non-em ones, the calibration constant of every sampling layer thus depends on the type of particle and on details of its shower development.

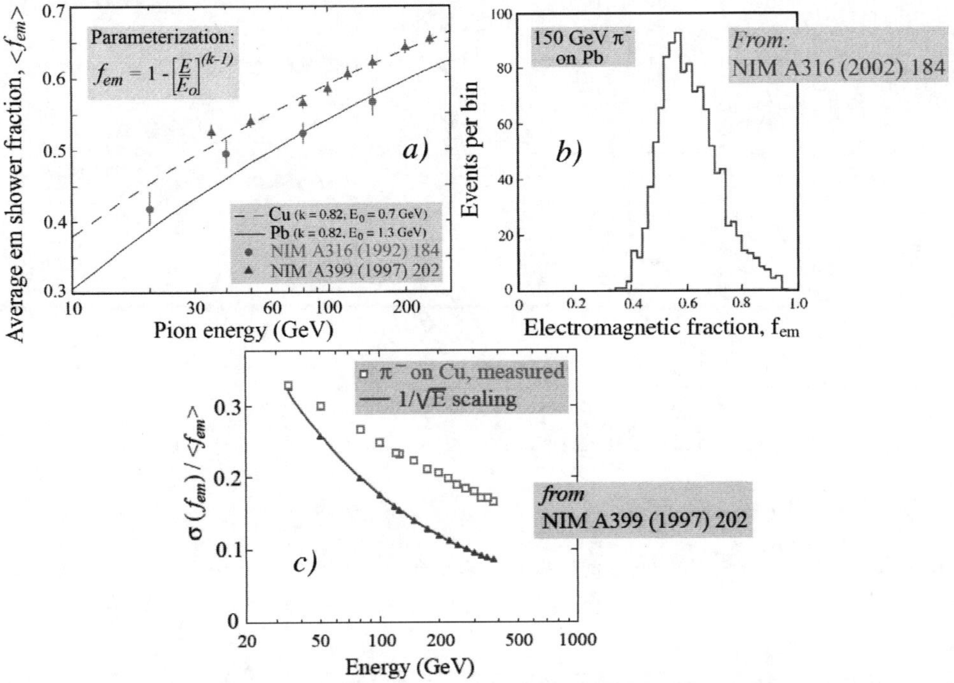

Figure 2. Effects of non-compensation on relevant hadronic calorimeter properties. Shown are the energy dependence of the average fraction of the hadron energy carried by the em shower component, $\langle f_{\rm em}\rangle$ (a), the event-to-event fluctuations in $f_{\rm em}$ (b), and the measured energy dependence of the rms width of the $f_{\rm em}$ distribution (c).

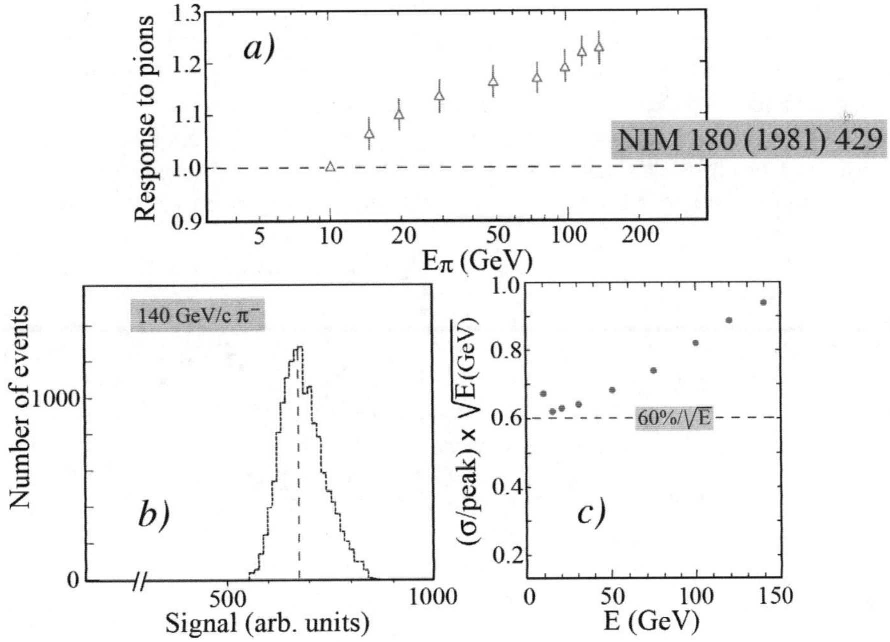

Figure 3. Experimental characteristics of a hadron calorimeter with $e/h = 1.5$. Shown are the hadronic signal non-linearity (a), the response function to 140 GeV/c pions (b) and the hadronic energy resolution as a function of energy (c). Data from [1].

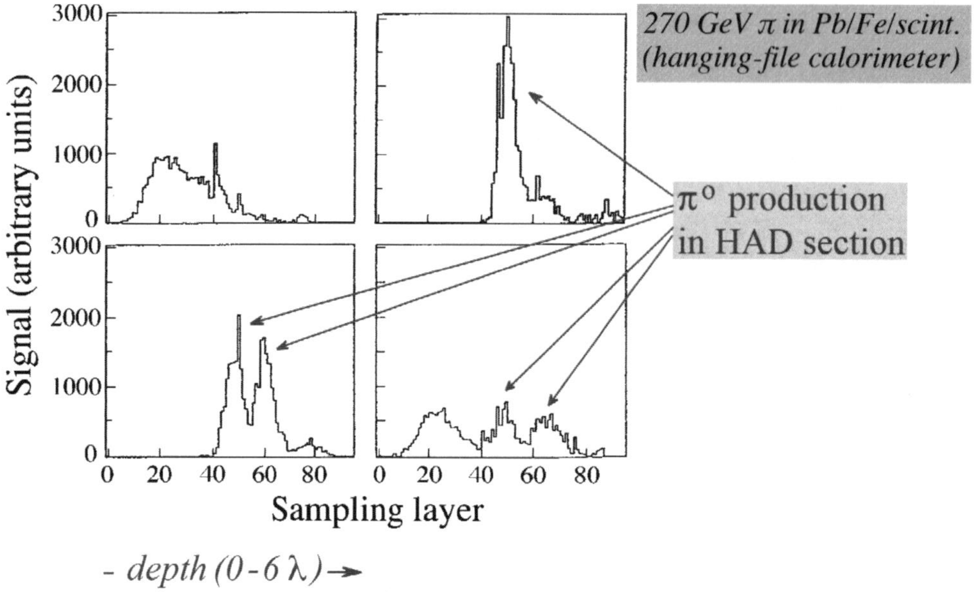

Figure 4. Longitudinal energy deposit profiles for 270 GeV π^- showers in a lead/iron/plastic- scintillator calorimeter. The 90 sampling layers, which cover a total depth of 6 nuclear interaction lengths, were read out individually in this experiment [3].

3. Calibration of longitudinally segmented calorimeter systems

Calibration problems of longitudinally segmented calorimeters are not limited to non-compensating hadron detectors. In fact, the earliest reports about this issue concern electron detection [4]. Let us, to illustrate the problem, consider a simple em calorimeter based on the detection of Čerenkov light, *e.g.,* a block of lead glass (Figure 5). When high-energy electrons or γs are absorbed, this detector produces, for example, 10 photoelectrons (p.e.) per GeV deposited energy. Therefore, 20 GeV electrons give a signal of 200 p.e., 100 GeV γs a signal of 1000 p.e., *etc.* Shower particles with energies below 0.3 MeV are not sufficiently relativistic to produce Čerenkov light and thus do not contribute to the signals. The relative contribution of such soft shower particles increases of course with the age of the shower, *i.e.,* with the depth inside the calorimeter. When this device is cut into three segments (I,II,III, see Figure 5), the number of photoelectrons produced per unit deposited energy is thus different for these

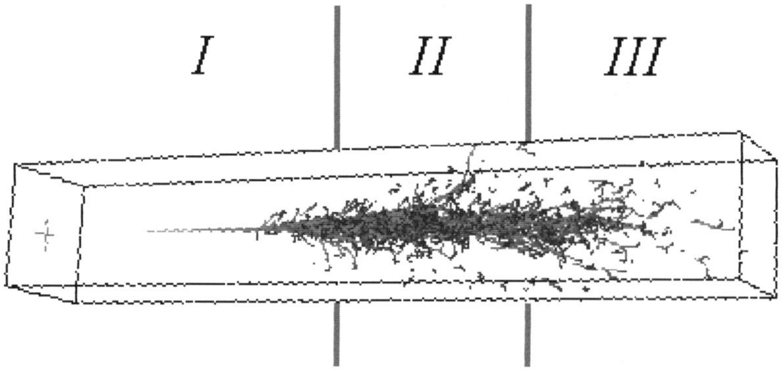

Figure 5. A high-energy electron shower developing in a lead glass calorimeter.

three segments, *e.g.,* 15, 10 and 5 p.e./GeV for the early (I), middle (II) and late (III) parts of a 100 GeV shower, which deposits typically 30, 40 and 30% of its energy in these 3 segments. As before, this yields a total signal of $100\% \times (0.3 \times 15 + 0.4 \times 10 + 0.3 \times 5) = 1000$ photoelectrons. In this detector, the number of photoelectrons per GeV constitutes the *calibration constant*, which relates the measured signal to the deposited energy. However, when these calibration constants (15 p.e./GeV *etc.*) are used for a 10 GeV electron, which deposits a larger fraction of its energy in segment I, the energy of this particle is systematically *overestimated.* On the other hand, the energy of a π^0 produced in segment III (calibration constant 5 p.e./GeV) by a hadron is systematically *underestimated.* The only way to avoid these problems is to use the calibration constant found for the unsegmented detector (10 p.e./GeV) for all three segments. However, in that case the energies deposited in these *individual* segments are not correctly determined from the measured signals. We conclude that **by cutting this calorimeter into longitudinal segments, problems are created that did not exist for the unsegmented detector**.

This example illustrates two of the major problems encountered when calibrating a longitudinally segmented calorimeter system: 1) Signal non-linearity, and 2) Dependence of the reconstructed energy on the starting point of the shower. The phenomenon that led to these problems in the example given above also plays an important role in all sampling calorimeters. The reason is that the composition of a developing shower changes and that the sampling fraction of the different shower particles contributing to the signal is not the same. As an example we mention em showers developing in a typical sampling

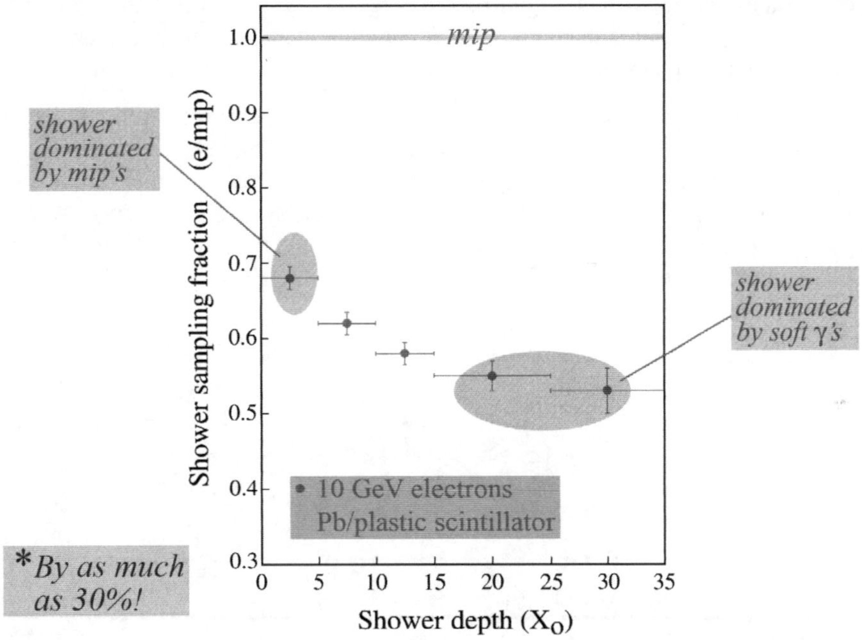

Figure 6. The sampling fraction as a function of the shower depth, or age, for 1 GeV electrons in a Pb/plastic scintillator calorimeter, normalized to the sampling fraction for a mip in this structure. Results from EGS4 Monte Carlo simulations.

calorimeter consisting of lead absorber and plastic scintillator active material (Figure 6). In such a structure, the abundantly produced soft γs ($E < 1$ MeV) are much less efficiently sampled than mips, because of the Z dependence of the photoelectric effect. As the shower develops, it changes gradually from a collection of particles dominated by mips to a collection of particles dominated by soft γs. The relationship between deposited energy and resulting signal decreases accordingly, by as much as 30%. This effect has caused major calibration problems for longitudinally segmented em calorimeters in ATLAS [5] and AMS [6]. These problems could only be resolved thanks to the availability of Monte Carlo simulations with which the mentioned issues and their consequences could be studied in great detail.

For hadron showers, the intercalibration problems are even more complicated than for em ones, because 1) em shower components may start at any point in the absorber structure, 2) the sampling fraction of the particles constituting the non-em shower component also changes with depth, 3) the calorimeter responses to the em and non-em components are not the same and 4) the energy sharing between these components varies wildly from event to event (Figure 4). Unfortunately, unlike for em showers, Monte Carlo simulations of hadron showers are not reliable enough to provide meaningful help for solving these problems. Just as in the case of em showers, where most of the energy carried by GeV-type particles is ultimately deposited in the last stages of the absorption process, through very soft electrons produced in photoelectric effect (with consequences for calorimetry shown in Figure 6), most of the energy carried by the non-em component is also deposited in the last stages of the absorption process, in this case through nuclear reactions. And as long as that part of the shower simulations is not correctly and completely implemented in the simulations, no reliable help in solving the complicated calibration problems should be expected. Experimental evidence for the strongly dominating contribution of soft protons to the non-em energy deposit is provided by

(i) The observation that calorimeters that are blind to such protons have very large e/h values. For example, in the DREAM Čerenkov calorimeter, an e/h value of 4.7 was measured, which means that only $\sim 20\%$ of the non-em energy was deposited by relativistic particles [7].

(ii) The absence of correlations between the signals from neighboring active layers in the fine-sampling ZEUS calorimeter. This proves that a typical shower particle contributing to the hadronic signals traveled less than the $0.06\lambda_{\rm int}$ separating the active layers [8].

Because of the absence of guidance from Monte Carlo simulations, and the emphasis in the design on the em energy resolution, the calorimeter systems of the LHC experiments face serious calibration challenges. This is particularly true for CMS, where the em and hadronic sections of the calorimeter have very different e/h values. The hadronic performance of this calorimeter system was systematically studied with various types of particles ($e, \pi, K, p, \bar{p}$), covering a momentum range from 1 - 300 GeV/c. Figure 7 shows some results from this study [9]. Both sections were calibrated with 50 GeV electrons.

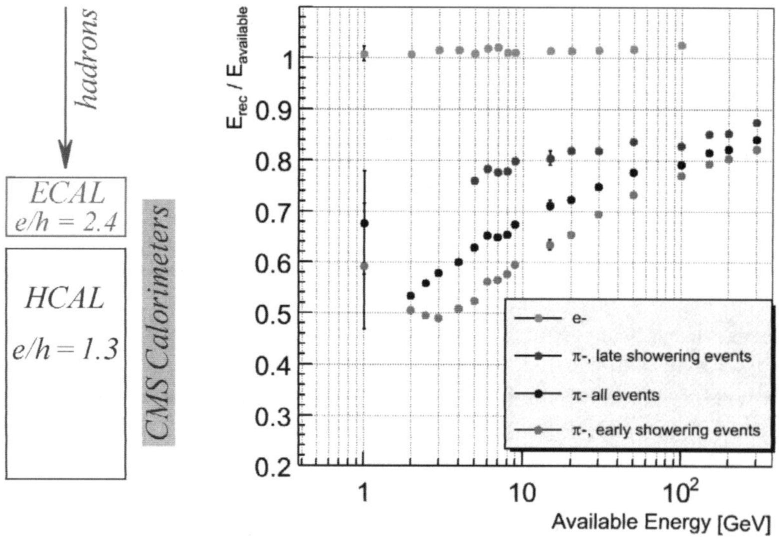

Figure 7. The response to electrons and pions as a function of energy, for the CMS barrel calorimeter. The pion events are subdivided into two samples according to the starting point of the shower, and the pion response is also shown separately for these two samples [9].

The figure shows that the calorimeter is very linear for electrons. The response to pions, represented by the black dots, indicates that the calorimeter is extremely non-linear for these particles. This non-linearity is especially evident below 10 GeV, which is important since pions in this energy range carry a large fraction of the energy of typical LHC jets, even in the TeV domain. More troublesome is the fact that the response strongly depends on the starting point of the showers. The figure shows results for two event samples, selected on that basis: showers starting in the em section (red) or in the hadronic section (blue). At low energies, the response is about 50% larger for the latter (penetrating) events. In practice in an experiment, it is often hard/impossible to determine where the shower starts, especially if these pions are traveling in close proximity to other jet fragments (*e.g.*, photons from π^0 decay) which develop showers in the em section.

Fortunately, the effects for jets are not as dramatic as Figure 7 may seem to suggest. This is because a jet consists typically of a large number of fragments, whose shower starting points fluctuate independently of each other. CMS have studied the jet response based on the very large database of events obtained in test beams of a variety of particles at a large number of energies, covering the range from 1 - 300 GeV. The response of the combined ECAL+HCAL system to electrons, pions, kaons, protons.

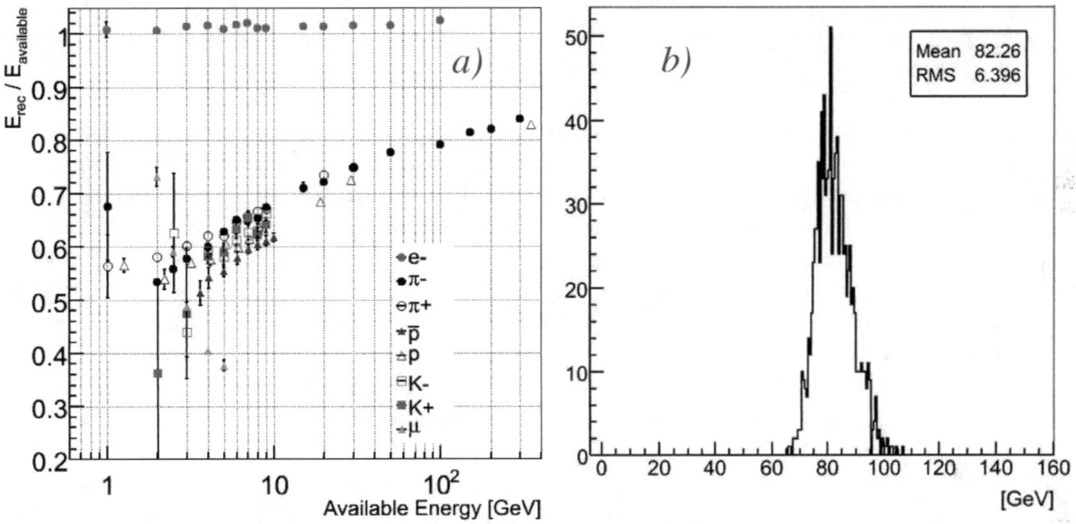

Figure 8. The response to different types of particles as a function of energy in the CMS barrel calorimeter (a), and the response function for one particular 100 GeV jet based on the measured response functions of the particles constituting this jet (b) [9].

antiprotons and muons is shown as a function of "available" energy[1] in Figure 8a. Differences between these particles observed in the for the jet response important energy range below 10 GeV are caused by differences in cross section, which affect the energy sharing between the two calorimeter sections. From the perspective of calorimetry, a jet is just a collection of particles, mainly pions and photons, which hit the calorimeter in the same region and develop simultaneous showers in it. The database summarized in Figure 8a can thus be used to reconstruct the response to a given jet in a semi-infinite number of ways. This is done in Figure 8b, which shows the response function to one particular 100 GeV jet, reconstructed 1000 times using measured events for the fragments of which it consisted, properly weighted to account for small differences in energy[2]. This procedure can be repeated for many different 100 GeV jets, and in

[1] Available energy denotes the energy that is available for producing a calorimeter signal. For protons, it is the kinetic energy, for mesons the kinetic energy plus the rest mass, and for antiprotons the kinetic energy plus twice the rest mass.

[2] For example, a 11.6 GeV π^+ generated in the fragmentation process was represented by an event from a 10 GeV π^+ run, and all calorimeter signals for this event were multiplied by a factor 1.16. Etcetera.

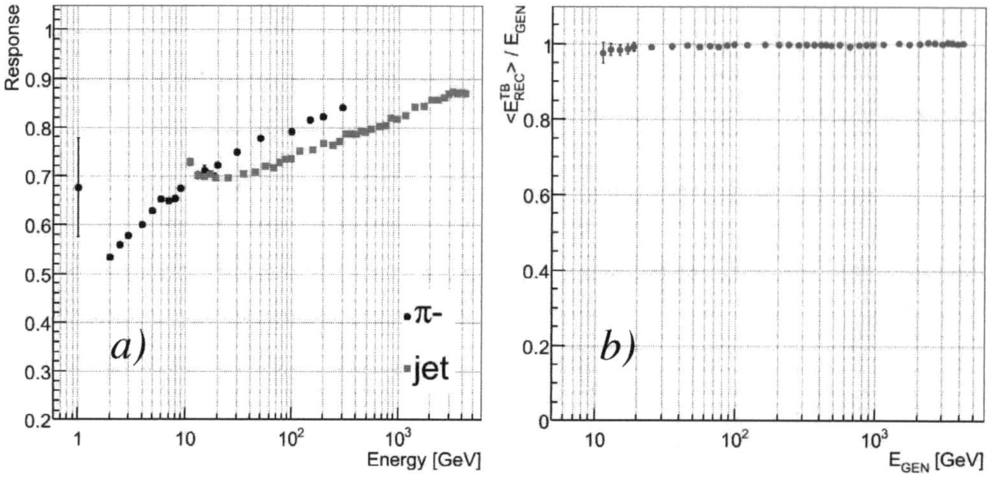

Figure 9. The CMS response to single π^- and to jets, as derived from the response functions of the constituting fragments(a). The reconstructed jet energy (normalized to the generated jet energy) after the inverse of the jet response is used as a correction factor (b). See text for details [9].

this way, the overall response function can be found for jets at this energy, or at any other energy. This is illustrated in Figure 9a, where the jet response obtained this way is given as a function of energy. For comparison, the response to single pions is shown as well. The fact that the jet response is generally smaller than the one to single pions of the same energy reflects the fact that the average hadronic jet fragment carries only a small fraction of the jet energy. The steeply decreasing response to low-energy pions apparently beats the fact that the response to the em jet fragments is larger (namely 1.0). Since the response is normalized to 1 for em showers, which were used to set the energy scale in both calorimeter sections in this analysis, the inverse of the jet response obtained in this way is the overall calibration constant with which the signals have to be multiplied in order to obtain the correct jet energy (Figure 9b). Figure 8b indicates that the energy resolution obtained for jets in this way is considerably better than the 14.8% measured for 100 GeV single pions.

Not all experiments set the scale in the different calorimeter sections in the same way, as in the example given above. In some experiments, more weight is given to the signals from the em section, in other methods the signals from the hadronic section are given a larger weight. The (lack of) merit of such methods has been extensively discussed elsewhere [10, 11, 12, 13]. Let me just comment that

(i) Such methods introduce unnecessary non-linearities and dependence on the starting point of the showers, and

(ii) Since the energy resolution is determined by event-to-event fluctuations and not by overall weighting factors, such methods offer no benefit whatsoever in this respect.

Ultimately the merits of the calibration procedures used by an experiment will be judged by how well a precisely known energy deposit is reproduced in practice. In em calorimetry, one can rely on abundantly produced resonances ($Z, \Upsilon, J/\psi \rightarrow e^+e^-$) for that purpose. Unfortunately, the situation is not that simple for the hadron calorimetry. Hadronically decaying intermediate vector bosons are an obvious candidate, but since UA2 no experiment has seen a W, Z bump in the minimum-bias jet-jet invariant mass distributions [14]. QCD background is blamed for that. However, given the larger cross sections, and the strongly improved b-tagging, one would expect the major experiments to have adequate samples of $Z \rightarrow b\bar{b}$ events, in which QCD background should be much less of a problem, for this purpose. Alternatively, one could possibly use samples of hadronically decaying Ws from t-quark decay or from diboson events. However, in all these cases integrated luminosities of several fb^{-1} would probably be needed for a meaningful check of the calibration procedures.

Of course, all calibration procedures of non-compensating hadron calorimeters are designed for "average" events. Jets of a specific type, *e.g.*, jets with a leading π^0, are likely to be reconstructed with an energy that is systematically wrong (too high in this case). Various methods are being employed to recognize such events and avoid this problem. Such methods are either based on the additional use of tracker information ("particle flow analysis"), or on the energy deposit pattern in the calorimeter itself ("offline compensation"), and are discussed in several other talks at this workshop.

Personally, I am in favor of longitudinally *unsegmented* calorimeters, for the following reasons:

(i) Calibration is trivial in that case, electrons in each tower and you are done. If there are no longitudinal segments, one is also not tempted to do the intercalibration wrongly.

(ii) As illustrated by the recent "spike" problem in the CMS ECAL, it is dangerous to place readout elements in the path of developing showers. As described in Section 2, most of the energy in the non-em component of hadron showers is deposited through nuclear reactions such as the one shown in figure 10. The nuclear fragments have dE/dx values that are typically 100-1000 times larger

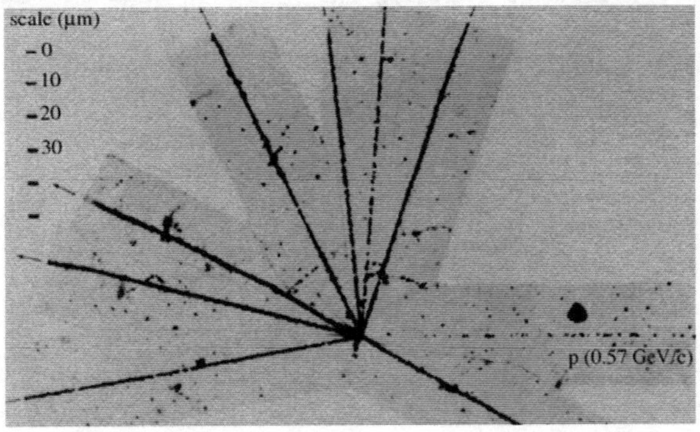

Figure 10. A typical nuclear reaction, through which non-em shower energy is deposited.

than for a mip. When traversing the APD that detects the scintillation light from the $PbWO_4$ crystals constituting the CMS ECAL, such (MeV type) particles may generate signals that are 100000 times larger that that from a scintillation photon, and thus fake energy deposits of tens of GeV.

(iii) And finally, there is not a single issue for which longitudinal segmentation is essential. In particular, electron identification can be accomplished in several excellent ways without the need for a separate em calorimeter section [15].

4. The future of calorimetry

It is expected that jet spectroscopy will become a major, if not the main, experimental technique at a future Lepton Collider. Separating hadronically decaying W and Z bosons is *the* benchmark for experiments at such a machine. The calorimeter systems currently operating at the high-energy frontier are not capable of this, as illustrated in Figure 11a, which shows the invariant jet-jet mass distribution for Ws and Zs reconstructed from jets in CMS as described in Section 3. The different approaches that are followed to develop calorimeter systems that are up to this task include:

- *Compensating* calorimeters. The reasons for the poor performance of non-compensating calorimeters, as well as the mechanisms to achieve compensation ($e/h = 1$), are completely

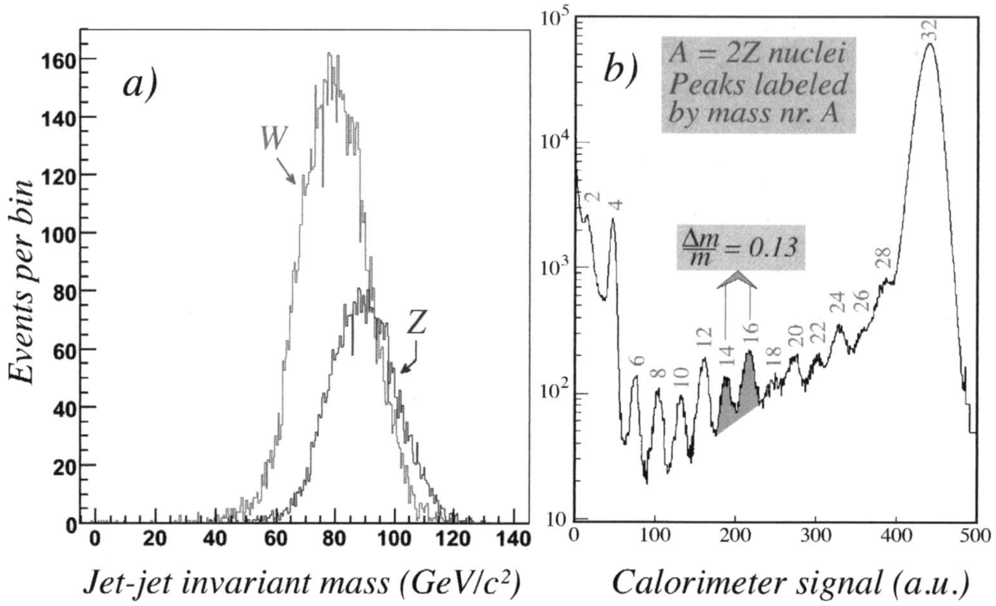

Figure 11. Reconstruction of the mass of hadronically decaying W, Z bosons from jets detected in the CMS calorimeters, using the method described in Section 3 (a). The compensating WA80 calorimeter as a high-resolution spectrometer (b) [16]. Note that the mass resolution needed to separate hadronically decaying W,Z bosons, $\Delta M/M \sim 0.11$.

understood since 25 years. All performance records for hadron calorimeters are held by the SPACAL [17] and ZEUS [18] calorimeters, which were designed based on this understanding. Figure 11b shows that the resolution obtained with such compensating calorimeters ($\sim 30\%/\sqrt{E}$) is good enough to turn them into high-precision jet spectrometers, with a resolving power that is adequate to separate hadronically decaying W and Z bosons. The compensation mechanism relies crucially on the detection of neutrons, which are abundantly produced in the nuclear reactions that take place in the hadronic absorption process. In order to boost the signal contribution of these neutrons to the required level, the sampling fraction has to have a very precise, small value (*e.g.,* 2.4% in the case of Pb/plastic-scintillator), which limits the em energy resolution that can be achieved with such calorimeters, to $\sim 13\%/\sqrt{E}$ in SPACAL. In addition, the signals have to be integrated over a rather large volume and time ($\sim$ 50 ns).

- *Dual-readout* calorimeters were invented to further *improve* on the excellent results obtained with compensating calorimeters, by avoiding the mentioned drawbacks of the latter. They are based on the idea that if the em fraction of hadronic showers ($f_{\rm em}$) can be measured event by event, the effects of fluctuations in this variable can be eliminated. The DREAM Collaboration has shown that this can be achieved by simultaneously measuring both the visible deposited energy and the Čerenkov light generated in the shower absorption process [7], exploring the fact that the latter is for all practical purposes only produced in the absorption of the em shower component. They have demonstrated that calorimeters based on this principle are linear for single hadrons and jets, have a Gaussian response function, and provide an energy resolution that scales with $E^{-1/2}$. The limitations of intrinsically compensating calorimeters do not apply to such devices, and excellent em energy resolution is not at all precluded, since one can even use the dual-readout principle for homogeneous crystals. Of course, it is in that case necessary to separate the light signals generated by such crystals into scintillation and Čerenkov components. The DREAM Collaboration has developed four different ways to achieve this [19, 20].
- Systems using *Particle Flow Analysis* are based on the combined use of a precision tracker and a

highly-granular calorimeter. The idea is that the charged jet particles can be precisely measured with the tracker, while the energy of the neutral particles is measured with the calorimeter. Such methods have indeed successfully been used to improve the mass resolution of hadronically decaying Z^0s at LEP, to $\sim$ 7 GeV/c^2 [21]. However, no one has ever come close to achieving the 3 - 3.5 GeV/c^2 resolution needed to separate W and Z bosons in this way. The problem that limits the success of this method is of course that the calorimeter does not know or care whether the particles it absorbs are electrically charged. Therefore, one will have to correct the calorimeter signals for the contributions of the charged jet particles. Proponents of this method have advocated a fine granularity as the key to the solution of this "double-counting" problem [22]. However, it has been argued by others that this, for practical geometries, is an illusion [23]. Especially in jets with leading charged particles, the overlap between the showers from individual jet particles makes the fine granularity largely irrelevant. In the absence of reliable Monte Carlo simulations, the only way to prove or disprove the advocated merits of the proposed methods is by means of dedicated experiments in realistic prototype studies. The CALICE Collaboration, which is building and testing detectors based on this principle, will hopefully make a significant contribution in this respect.

5. Concluding remarks

Interpretation of the calorimeter signals is crucially important for the jet energy scale. The need to (inter)calibrate the various segments of a longitudinally subdivided calorimeter system makes this a very complicated issue, especially when the segments have different e/h values. Naively, one would think that the additional information provided by longitudinal segmentation would allow one to make more precise measurements. However, I think this is an illusion. There is in that respect an analogy with the calorimetry practiced in thermodynamical experiments. Just as it does not help to know the 4-vectors of all the molecules when one wants to measure the specific heat of a certain substance, the additional information provided by the longitudinal calorimeter segments mainly complicates and confuses the measurement of the energy deposited in the absorption of jets.

The absence of reliable Monte Carlo simulations is becoming more and more problematic as the need for high-precision jet spectroscopy increases. In Section 3, we saw how simulations were crucial for resolving complicated calibration problems encountered with the electromagnetic calorimeters of ATLAS and AMS. I think that the lack of equivalently useful tools for hadron absorption in matter has actually led to a *decrease* of the quality of hadron calorimetry in the experiments developed in the last 20 years. Unfortunately, as long as those who could make a difference in that respect are not aware or, worse, in denial of the existing problems, this is unlikely to change.

References

[1] Abramowicz H *et al.* 1981, Nucl. Instr. and Meth. **180**, 429.
[2] Akchurin N *et al.* 1998, Nucl. Instr. and Meth. **A408**, 380.
[3] Green D 1994, *Proc. 4th Int. Conf. on Calorimetry in High Energy Physics*, La Biodola, Italy, eds. A. Menzione and A. Scribano, (Singapore: World Scientific), p. 1.
[4] Åkesson T *et al.* 1987, Nucl. Instr. and Meth. **A262**, 243.
[5] Aharrouche M *et al.* 2006, Nucl. Instr. and Meth. **A568**, 601.
[6] Cervelli F *et al.* 2002, Nucl. Instr. and Meth. **A490**, 132.
[7] Akchurin N *et al.* 2005, Nucl. Instr. and Meth. **A537**, 537.
[8] Drews G *et al.* 1990, Nucl. Instr. and Meth. **A290**, 335.
[9] Akchurin N *et al.* 2007, *The Response of the CMS Combined Calorimeters toSingle Hadrons, Electrons and Muons*, CMS Note 2007/012, CERN, Genève, Switzerland.
[10] Ganel O and Wigmans R 1998, Nucl. Instr. and Meth. **A409**, 621.
[11] Wigmans R and Zeyrek M 2002, Nucl. Instr. and Meth. **A485**, 385.
[12] Albrow M *et al.* 2002, Nucl. Instr. and Meth. **A487**, 381.

[13] Wigmans R 2006, *On the Calibration of Segmented Calorimeters*, Proc. XII Int. Conf. on Calorimetry in High Energy Physics, Chicago 2006, AIP Conf. Proc. **867**, 90
[14] Jenni P 1988, *Nucl. Phys. B (Proc. Suppl.)* **3**, 341.
[15] Wigmans R 2011, *Calorimeters*, in Handbook of Particle Detection and Imaging, ed. C. Grupen, Springer Verlag (2011).
[16] Young GR *et al.* 1989, Nucl. Instr. and Meth. **A279**, 503.
[17] Acosta D *et al.* 1991, Nucl. Instr. and Meth. **A308**, 481.
[18] Behrens U *et al.* 1990, Nucl. Instr. and Meth. **A289**, 115.
[19] Akchurin N *et al.* 2009, Nucl. Instr. and Meth. **A604**, 512.
[20] Akchurin N *et al.* 2011, Nucl. Instr. and Meth. **A638**, 47.
[21] Buskulic D etal 1995, Nucl. Instr. and Meth. **A360**, 481.
[22] *TESLA Technical Design Report* 2001, report DESY 2001-011, DESY, Hamburg, Germany.
[23] Lobban O, Sriharan A and Wigmans R 2002, Nucl. Instr. and Meth. **A495**, 107.

Jet measurements in ATLAS

Peter Loch (on behalf of the ATLAS Collaboration)

Department of Physics, University of Arizona, Tucson, Arizona 85721, USA

E-mail: loch@physics.arizona.edu

Abstract. The reconstruction of jets generated in the proton-proton collisions at the Large Hadron Collider (LHC) at a center of mass energy of $\sqrt{s} = 7$ TeV with the ATLAS detector is discussed. Beginning with a brief review of the calorimeter signal definitions relevant for jet finding, and the use of reconstructed charged particle tracks, the jet reconstruction strategy is described in some detail. Emphasis is put on the jet energy scale (JES) calibration strategy applied for first data, which is based on a short sequence of data driven and simulation based calibrations and corrections to restore the measured jet energy to particle level. The level of understanding of the signal patterns entering the JES corrections is shown for selected variables in comparisons to simulations. The present systematic uncertainties on the JES, which can be as low as 2% for central jets, are presented and analyzed with respect to the individual fractional contributions entering their determination. Some characteristic jet reconstruction performance and selected results from the first year of jet physics with ATLAS in a newly accessible kinematic domain are shown in conclusion.

1. Introduction

The ATLAS experiment [1] at the Large Hadron Collider (LHC) commenced its first successful year of proton physics running at $\sqrt{s} = 7$ TeV in early November 2010. A total of about 45 fb^{-1} of proton-proton (pp) collision data were recorded, with instantaneous luminosities peaking at approximately 2×10^{32} $\text{cm}^{-2}\text{s}^{-1}$, thus generating up to about 3.5 pp collisions per LHC bunch crossing on average.

Collimated bundles of particles (jets) in the final state of a pp collision are not only produced by the hadronization of high energy partons emerging from the hardest scattering process, but also by initial and final state (gluon) radiation associated with this scattering, and in the softer multiple parton interactions in both the underlying event and the possible additional pp collisions in the same proton bunch crossing (pile-up). Any of these jets with transverse momentum (p_{T}) of more than a few GeV leave significant signals in the ATLAS detector system and are reconstructed for the characterization and measurement of the actual final state. The quality of this reconstruction and the calibration strategies applied in its context are discussed in this note, together with selected examples of the reconstruction performance and jet physics measurements. The discussion of the jet calibration strategies is intentionally restricted to the methods applied in the default ATLAS jet reconstruction for the 2010 pp data, which has been used for all published jet reconstruction performance evaluations, jet physics analysis and searches for new physics so far. Specifically, contributions to this workshop describing some of these results include in-situ jet energy scale validations [2], inclusive jet and di-jet cross sections [3], jets in $t\bar{t}$ final states [4] and jets in searches for new physics [5].

The structure of this note is as follows. In Section 2, the ATLAS experiment is briefly described. The detector signals used for the reconstruction of particle jets and other physics objects are discussed in Section 3, while the jet reconstruction and calibration sequences are presented in Section 4. The determination the systematic jet energy scale uncertainties and their fractional contributions are shown in Section 5, while some results from jet physics sensitive to the jet reconstruction quality are discussed in Section 6. This note concludes with a brief summary and outlook in Section 7.

2. The ATLAS experiment

The ATLAS collaboration designed, built and now operates one of two multi-purpose experiments at the LHC. Its detector system features the inner detector (ID) tracking system closest to the collision vertex in radius, which provides charged particle track reconstruction for pseudorapidities $|\eta| < 2.5$ in a solenoid field of 2 T. The ID contains from inside out a silicon pixel detector, a silicon strip detector for high precision track and vertex reconstruction, and a transition radiation detector featuring a large number of straw tubes arranged in a large number of planes (from 73 central up to 160 in the end cap) for large radius tracking.

The ID is surrounded by electromagnetic calorimetry covering $|\eta| < 3.2$, featuring liquid argon calorimeters with lead absorbers folded into an accordion ($|\eta| < 1.4$) or a "Spanish fan" shape ($1.4 < |\eta| < 3.2$). These calorimeters provide high precision reconstruction for electrons and photons without any azimuthal discontinuities. The readout geometry of these devices is projective with respect to the nominal collision vertex in ATLAS, and highly segmented, with up to three longitudinal layers and a typical radial granularity in η and azimuth (ϕ) of $\Delta\eta \times \Delta\phi = 0.025 \times 0.025$. The depth of the electromagnetic calorimeters is between $22 - 26$ radiation length (X_0) and up to 1.5 hadronic interaction length (λ). The total number of readout channels is about $170,000$.

Hadronic calorimetry in the central region of ATLAS is provided by a scintillator tile/iron calorimeter surrounding the electromagnetic calorimetry with coverage up to about $|\eta| < 1.7$. The typical readout granularity is $\Delta\eta \times \Delta\phi = 0.1 \times 0.1$ in three longitudinal layers. Liquid argon/copper calorimetry featuring a parallel plate absorber structure with four longitudinal segments and the same projective granularity covers $1.5 < |\eta| < 3.2$ at both ends of the central calorimeters. The gap between these end-cap calorimeters and the LHC beam line is closed by three modules of forward calorimeters at both ends of the detector, covering $3.2 < |\eta| < 4.9$. All forward calorimeters are high density liquid argon devices with a large number of tubular electrodes arranged parallel to the beam line. The modules closest to the collision vertex feature a copper absorber, while the modules behind it have a tungsten absorber. Overall, ATLAS calorimetry provides nearly continuous hadronic coverage of at least 10 λ within $|\eta| < 4.9$, with a total of nearly $190,000$ independently read out cells. The ATLAS calorimeters are non-compensating, with a typical e/π signal ratio of about 1.3.

The ATLAS calorimeters are surrounded by a muon spectrometer covering $|\eta| < 2.7$ and featuring muon chambers in an air core toroid magnetic field, which allows high precision reconstruction of all three components of the muon momentum. The overall dimensions of the experiment including the muon system are about 46 m long and 23 m high, at a total weight of approximately $7,000$ t. A comprehensive overview of all detector systems can be found in Ref. [1].

3. Input signals

The most important detectors for jet reconstruction in ATLAS are the calorimeters. The performance of the individual systems has been evaluated in extensive test beam programs for the various modules, see e.g. Refs. [6–10], and in-situ, see for example Refs. [11, 12].

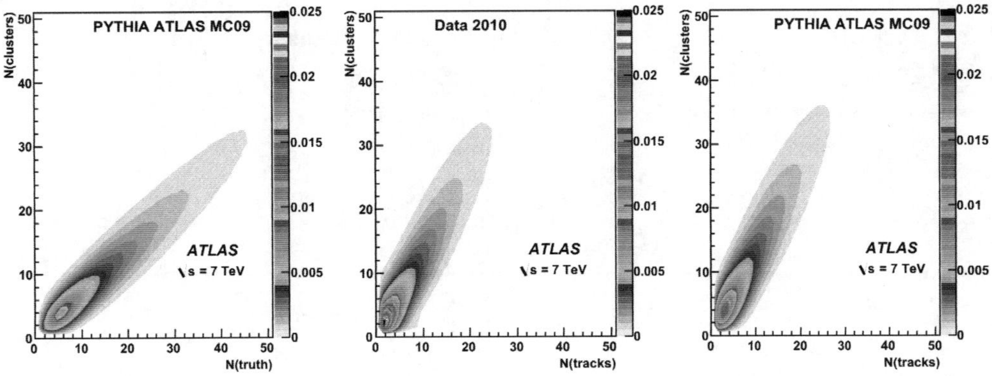

Figure 1. The correlation between the number of topological cell clusters and the number of incoming particles, as reconstructed from fully simulated *pp* collision events in ATLAS (left). The correlation between the number of clusters and the number of reconstructed charged tracks in data is shown in the center, while the same correlation in full simulation is shown on the right (figure from Ref. [13]).

3.1. Calorimeter towers and cluster

The about 190,000 cell signals collected in each recorded collision event by the ATLAS calorimeters are reconstructed on a basic electromagnetic energy scale, which has been determined for each module in the system using electron test beam and simulations. The cell signals are further collected into calorimeter towers or topological cell clusters. In case of calorimeter towers, the event energy flow is projected onto a regular $\Delta\eta \times \Delta\phi = 0.1 \times 0.1$ grid within the detector acceptance ($-\pi < \phi \leq \pi$, $-5 < \eta < 5$). The cell signal contribution to a tower in this grid[1] is determined by the overlap of the cell area in (η, ϕ) space with the tower. If a projective cell is smaller than or equal to the tower bin size, its energy E_{cell} is added in full to the energy in the tower it overlaps. Larger projective cells contribute $w_{\text{cell}}E_{\text{cell}}$ to any tower they overlap, with the geometrical weight $w_{\text{cell}} < 1$ reflecting the sharing of E_{cell} according to the fraction of area shared between tower and cell. Non-projective cells like the ones in the forward calorimeters can be smaller than the tower, yet distribute their energy among several bins with w_{cell} depending on the projection of the actual cell front face onto the tower grid.

Topological cell clusters [14] are three dimensional signal objects formed from cells following significant spatial energy flow patterns in the calorimeters. The main motivation is the attempt to reconstruct individual particle showers in the calorimeter, or at least provide a measure highly correlated with the spatial energy flow in the event even if particle bundles are too collimated (like in jets) to distinguish individual showers within the calorimeter readout granularity. Figure 1 from Ref. [13] indicates this correlation between the multiplicities in particle flow and clusters.

The cluster algorithm starts from cells whose signal is more than $4\sigma_{\text{noise}}$, around which signals from directly neighboring cells are collected in all three dimensions. If any of these signal is more than $2\sigma_{\text{noise}}$, its neighbors are collected as well, independent of their signal significance. Finally, the collected cluster is split if it contains more than one local signal maximum. This clustering approach produces "energy blobs" around significant signals in the calorimeter, which can cross sub-system boundaries and provide shape and location information. The algorithm also acts as a noise suppression tool, as cells with insignificant signals away from seeds are not included into any topological cluster and dropped from jet and other physics object reconstruction which uses

[1] The direction $(\eta_{\text{cell}}, \phi_{\text{cell}})$ of the cell is defined by the nominal collision vertex and the cell location.

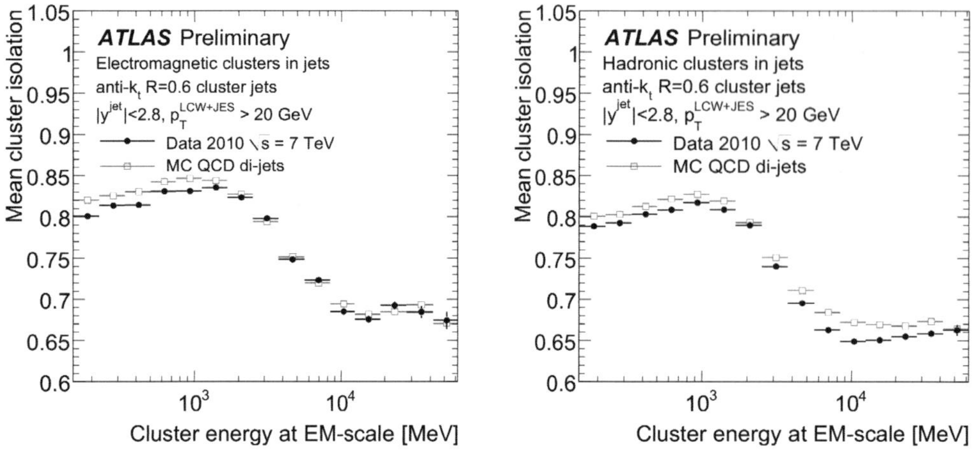

(a) Electromagnetic clusters

(b) Hadronic clusters

ge isolation of topological cell clusters in jets. for electromagnetic and unction of the cluster energy on electromagnetic scale. Full simulations ompared to data (from Ref. [16]).

rviving cells in clusters can also be collected into the already discussed ıs creating a noise-suppressed tower signal.

ıy scales

above, the calorimeter cell signal E_{cell} in ATLAS is reconstructed on the energy scale. Calorimeter towers contain sums of possibly weighted cell ower energy E_{tower} is also reconstructed on this basic scale. In addition, tion $(\eta_{\mathrm{tower}}, \phi_{\mathrm{tower}})$ which is defined by its bin center in the (η, ϕ) grid. our-momentum $(E_{\mathrm{tower}}, \vec{p}_{\mathrm{tower}})$, with $E_{\mathrm{tower}} = |\vec{p}_{\mathrm{tower}}|$, is constructed for l other further reconstruction. Topological cell clusters provide additional their total energy on electromagnetic scale, which is the sum of cell signals heir location and shape hints to the origin of the signal, which for high l more or less completely in the electromagnetic calorimeter is very likely Larger clusters and/or clusters located deep inside the calorimeters are n. These features provide a basis for a local hadronic calibration, where fines the calibration and correction functions for both electromagnetic For hadronic clusters, a non-compensation correction is applied to each ll signal weighting approach), depending on the cluster location, energy, nsity. Additional corrections are then applied to recover out-of-cluster energy loss in nearby inactive material. The last two are also applied to rs, with different parameters reflecting the differences in the underlying Like towers, clusters are represented by a massless four-momentum, either hadronic energy scale.

calibrations are derived independent from each other using detailed single ased on the Geant4 toolkit [17, 18]. Each correction uses only cluster ding on any other correction, and, even though the effects of e.g. a poor ıd the amount of energy lost outside of clusters or in inactive material are hadronic shower, the correction functions and parameters at cluster level are not. It should also be noted that all calibrations and corrections exploit the available knowledge of the relation between deposited energy and signal in the detector simulation, and are not results of numerical fits.

The completely simulation based determination of the calibrations and corrections for the local hadronic energy scale requires good modeling of the observable parameters used in the various procedures. For example, the out-of-cluster corrections derived for single particles are topology dependent, as energy lost for one cluster may generate a signal in a nearby cluster. To scale this correction for a jet environment, the spatial cluster isolation is explicitly measured and used in this correction. Figure 2 from Ref. [16] shows a comparison of the average isolation of clusters in jets as a function of the cluster energy. Simulations follow the data very well with this complex observable.

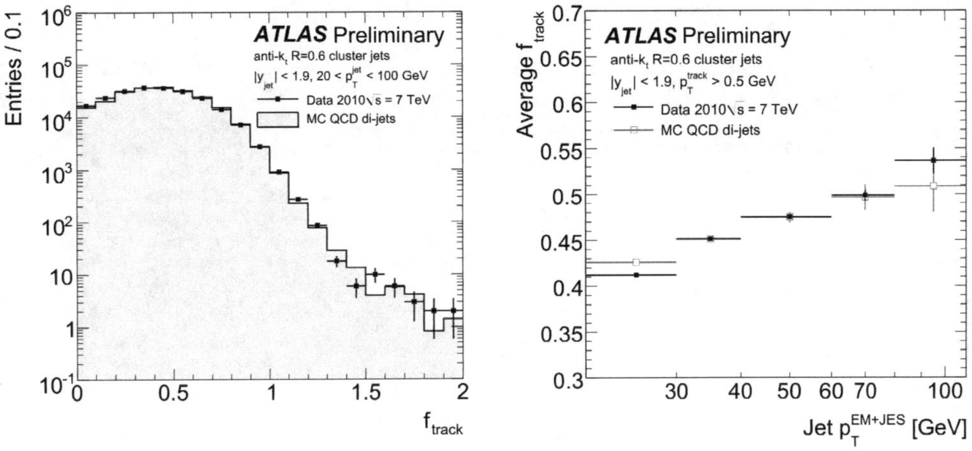

Figure 3. The spectrum of the charged track p_{T} ratio f_{track}, as described in the text, in comparison of data with simulations (left). The scaling of $\langle f_{\mathrm{track}} \rangle$ with jet p_{T} is shown in the right figure, again for data and simulations (taken from Ref. [16]).

3.3. Reconstructed tracks

In addition to the calorimeter signals, ATLAS also uses tracks to reconstruct jets, which provide independent detection of jet activity and measurements of jet properties like width. Tracks pointing to calorimeter jets can also be used to refine jet calibration. As for calorimeter signals, good modeling of basic track properties is essential. One of the characteristic parameters considered for calibration refinementis the ratio of charged particle transverse momentum ($p_{\mathrm{T}}^{\mathrm{track}}$) to the total measured p_{T} from the calorimeter, $f_{\mathrm{track}} = p_{\mathrm{T}}^{\mathrm{track}}/p_{\mathrm{T}}$, depicted in Fig. 3. Simulations reproduce the f_{track} spectrum and scaling of the average f_{track} with p_{T} very well.

4. Jet reconstruction and calibration

The ATLAS jet reconstruction implementation supports calorimeter towers and clusters, and reconstructed tracks, as input signals, in addition to simulated particles from physics generators. The corresponding software is designed to allow use of all jet finding strategies of interest on any of these inputs, with only a few input specific tools like those for jet calibration to be configured. The jet algorithm suite provided by FastJet [19] and the infrared safe seedless cone algorithm SISCone [20] are used within this framework. The present default jet definition used by ATLAS

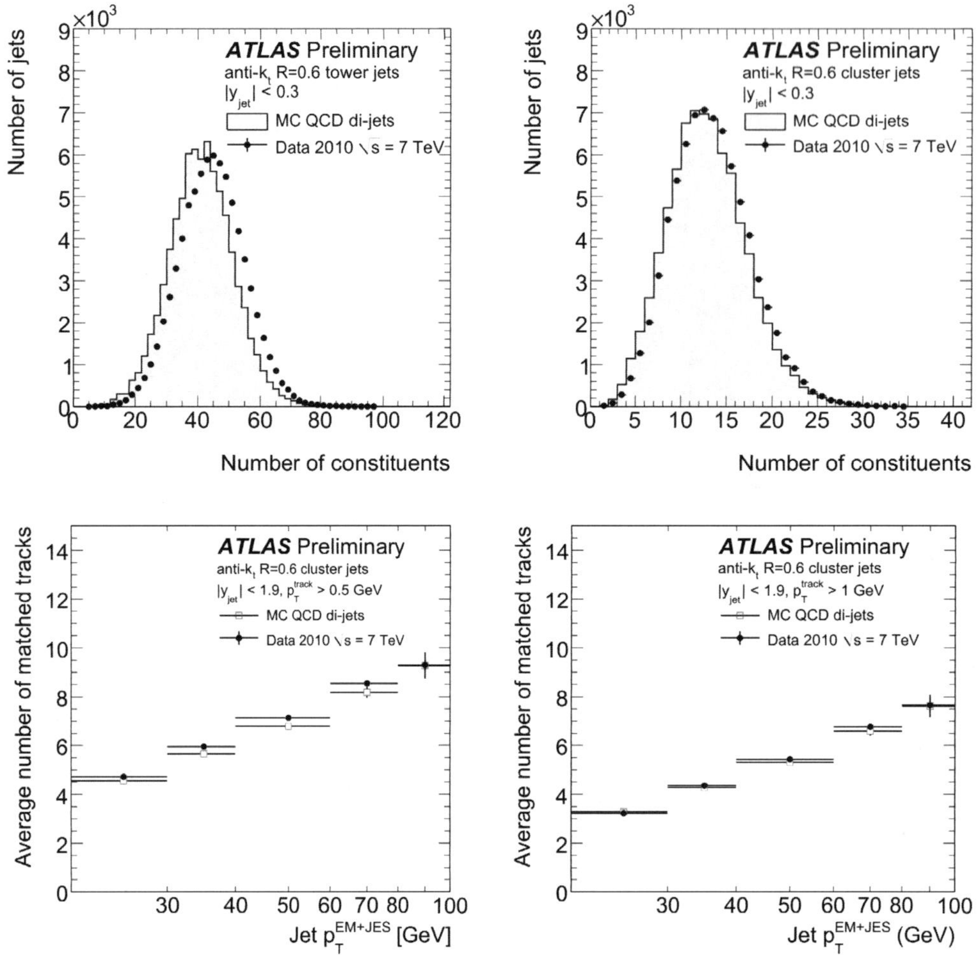

Figure 4. The number of constituents in wide central jets reconstructed with the ATLAS calorimeters (top left from calorimeter towers, top right from topological cell clusters). The scaling of the average number of tracks pointing to the cluster jets as function of jet p_T is shown at the bottom (left, including tracks with $p_T > 500$ MeV, right for tracks with $p_T > 1$ GeV). Data is shown in comparison to simulations (from Ref. [16]).

is the anti-k_T algorithm [21] with full four-momentum recombination ("E-scheme") and two jet sizes, $R = 0.4$ and $R = 0.6$.

4.1. Jet formation

The jet reconstruction sequence starts with a list of calorimeter towers, topological cell clusters, reconstructed tracks, or generated particles. In the case of clusters and towers, all objects with $E_{\text{tower}} < 0$ or $E_{\text{cluster}} < 0$ are rejected[2], to ensure meaningful four-momentum algebra during jet formation. In the present default configuration, the calorimeter signals enter jet finding at the electromagnetic energy scale - the use of locally calibrated clusters is under commissioning.

[2] Intentionally, negative energy tower and cluster signals are not generally removed from the calorimeter signal; rather, they are used to e.g. monitor noise characteristics.

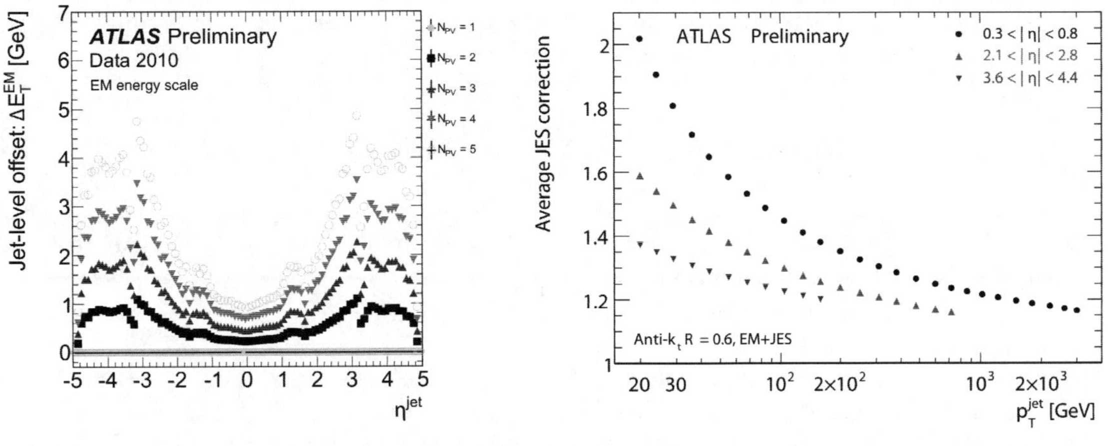

Figure 5. The average pile-up correction $\Delta E_{\mathrm{T}}(N_{\mathrm{PV}}, \eta) = \langle \rho_{\mathrm{T}} \rangle (N_{\mathrm{PV}}, \eta) \times A_{\mathrm{jet}}(R)$ for $R = 0.6$ anti-k_{T} jets (from Ref. [22]) is shown in the left figure. The jet energy scale correction $\mathcal{R}_{\mathrm{em}}^{-1}(E_{\mathrm{em+PU}}, \eta)$ is shown as function of the fully corrected jet p_{T} in the right figure, taken from Ref. [23].

The configured jet clustering algorithm of choice is then applied to the selected (now physically meaningful) input objects, forming jets with links back to their constituents. This allows the measurement of shapes and other jet characteristics, and the analysis of jet substructure. The number of constituents forming a jet naturally depends not only on its particular fragmentation, but also on the input signal definition in case of jets reconstructed with the calorimeter (calorimeter jets), and the track selection cuts for tracks in jets, see Fig. 4 from Ref [16].

4.2. Jet calibration

The present calibration scheme for calorimeter jets in ATLAS establishes a flat response to jets within the full coverage of the calorimeters. The approach features few correction levels, thus limiting the algorithm biases and reducing sensitivity to modeling details in the simulation. It is based on the electromagnetic energy scale signal of jets, given by (example for cluster jets, similar for towers):

$$(E_{\mathrm{em}}, \vec{p}_{\mathrm{em}})_{\mathrm{jet}} = \sum_{i=1}^{N_{\mathrm{cluster}}^{\mathrm{jet}}} (E_{\mathrm{em}}, \vec{p}_{\mathrm{em}})_{\mathrm{cluster},i} \quad \text{with } E_{\mathrm{em}} > 0 \text{ for all cluster } i \qquad (1)$$

E_{em} and $\vec{p}_{\mathrm{em}}$ are the energy and momentum reconstructed on the electromagnetic energy scale, for jets and clusters. $N_{\mathrm{cluster}}^{\mathrm{jet}}$ is the number of clusters forming the jet. The jet level calibration sequence then is

(1) Pile-up correction (derived from data) As already mentioned, with increasing LHC luminosity the number of additional pp collisions in same bunch crossing as the triggered hard scattering increases, thus adding energy to the event. These extra collisions are completely incoherent with the hard scatter and each other, and the added energies are mostly from soft QCD processes similar to those measured in minimum bias events. They give rise to a transverse energy density $\rho_{\mathrm{T}}(N_{\mathrm{PV}}, \eta) = \langle E_{\mathrm{T}} \rangle (N_{\mathrm{PV}}, \eta)/(\Delta\eta \times \Delta\phi)$, which is measured on the electromagnetic energy scale with minimum bias events taken at the same luminosities. N_{PV} is the number of reconstructed primary vertices in the event, and as such a measure of the number of additional collisions. The corrected jet momentum for a jet with reconstructed direction η_{jet} is then

$$\begin{aligned}(E_{\mathrm{em+PU}}, \vec{p}_{\mathrm{em+PU}})_{\mathrm{jet}} &= \\ &(E_{\mathrm{em}} - \Delta E_{\mathrm{T}}(N_{\mathrm{PV}}, \eta_{\mathrm{jet}}) \cosh \eta_{\mathrm{jet}}, (1 - \Delta E_{\mathrm{T}}(N_{\mathrm{PV}}, \eta_{\mathrm{jet}}) \cosh \eta_{\mathrm{jet}}/E_{\mathrm{em}}) \cdot \vec{p}_{\mathrm{em}})_{\mathrm{jet}},\end{aligned}$$

with $\Delta E_{\mathrm{T}}(N_{\mathrm{PV}}, \eta_{\mathrm{jet}}) = \rho_{\mathrm{T}}(N_{\mathrm{PV}}, \eta_{\mathrm{jet}}) \cdot A_{\mathrm{jet}}^{\eta\phi}$ and the jet area $A_{\mathrm{jet}}^{\eta\phi}$ measured in (η, ϕ) space. Figure 5 (left) shows the average amount of pile-up energy for anti-k_{T} jets with $R = 0.6$ with the 2010 LHC beam conditions.

(2) Event vertex correction (from event data) Calorimeter jets are reconstructed from clusters or towers located in the nominal geometry reference frame of ATLAS, with the central vertex defining the experiment center assumed to be located at $\vec{x}_{\mathrm{center}} = (0., 0., 0.)$. On the other hand, not only is the LHC beam spot systematically shifted with respect to $\vec{x}_{\mathrm{center}}$, the actual primary event vertex $\vec{x}_{\mathrm{PV}}$ is also reconstructed by the ID. The momentum $\vec{p}_{\mathrm{em+PU}}$ of each calorimeter jet, and of all clusters forming this jet, is corrected for the change of directions induced by this vertex shift ($\vec{p}_{\mathrm{em+PU}} \rightarrow \vec{p}_{\mathrm{em+PU+PV}}$), while the jet energy is unchanged. The new jet four momentum is then $(E_{\mathrm{em+PU}}, \vec{p}_{\mathrm{em+PU+PV}})$.

(3) Response correction (from simulations) The paramount correction in the present scheme is the response correction, where signal inefficiencies due to the non-compensation, signal definition, crack regions in the calorimeters, energy losses in inactive material, and the charged particle energy losses due to the magnetic solenoid field in front of the calorimeter are corrected. The correction is determined using QCD di-jet events generated by Pythia and processed through the full simulation of the ATLAS detector to model detector signals. The simulated calorimeter jets are matched with "true" generated particle jets built with the same anti-k_{T} algorithm configuration. The total energy of the particle jets $E_{\mathrm{jet}}^{\mathrm{true}}$ is then compared to the reconstructed matched jet energy $E_{\mathrm{em+PU,jet}}$ (see item (1) above) in regions defined by the reconstructed jet direction η_{jet} and as function of $E_{\mathrm{jet}}^{\mathrm{true}}$ and denoted as response $\mathcal{R}_{\mathrm{em}}(E_{\mathrm{jet}}^{\mathrm{true}}, \eta_{\mathrm{jet}}) = \left\langle E_{\mathrm{em+PU,jet}}/E_{\mathrm{jet}}^{\mathrm{true}} \right\rangle (E_{\mathrm{jet}}^{\mathrm{true}}, \eta_{\mathrm{jet}})$. Applying the numerical inversion technique described in Ref. [24] allows to change from a truth energy to a measured energy parametrization, i.e. $\mathcal{R}_{\mathrm{em}}(E_{\mathrm{jet}}^{\mathrm{true}}, \eta_{\mathrm{jet}}) \rightarrow \mathcal{R}_{\mathrm{em}}(E_{\mathrm{em+PU,jet}}, \eta_{\mathrm{jet}})$ while maintaining the shape of the response function and avoiding calorimeter energy resolution biases. The response correction is applied as follows:

$$(E_{\mathrm{em+PU+R}}, \vec{p}_{\mathrm{em+PU+PV+R}})_{\mathrm{jet}} = (E_{\mathrm{em+PU}}, \vec{p}_{\mathrm{em+PU+PV}})_{\mathrm{jet}} \times \mathcal{R}_{\mathrm{em}}^{-1}(E_{\mathrm{em+PU,jet}}, \eta_{\mathrm{jet}}).$$

The response corrections can be significant, as indicated in Fig. 5 (right). It decreases for more forward going jets, as expected from the increasing jet energies (at a fixed p_{T}) in this region.

(4) Direction correction (from simulations) The direction η_{jet} of jets close to poorly instrumented regions of the ATLAS calorimeters, like transitions between calorimeter modules and cracks between cryostat vessels, is often biased with respect to the true jet direction, i.e. depending on the shape of the problem region it may be pulled forward towards the beam or pulled back towards the central region. The corresponding small shifts in η have been computed using simulations and are used for a direction correction $\Delta\eta_{\mathrm{jet}}(E_{\mathrm{em+PU}}, \eta_{\mathrm{jet}})$, parametrized as a function of the jet energy and the original jet direction (no vertex correction applied). This correction works only on the jet momentum and leaves the energy unchanged:

$$(E_{\mathrm{em+JES}}, \vec{p}_{\mathrm{em+JES}})_{\mathrm{jet}} = (E_{\mathrm{em+PU+R}}, \vec{p}_{\mathrm{em+PU+PV+R}} \otimes \Delta\eta_{\mathrm{jet}}(E_{\mathrm{em+PU}}, \eta_{\mathrm{jet}}))\,.$$

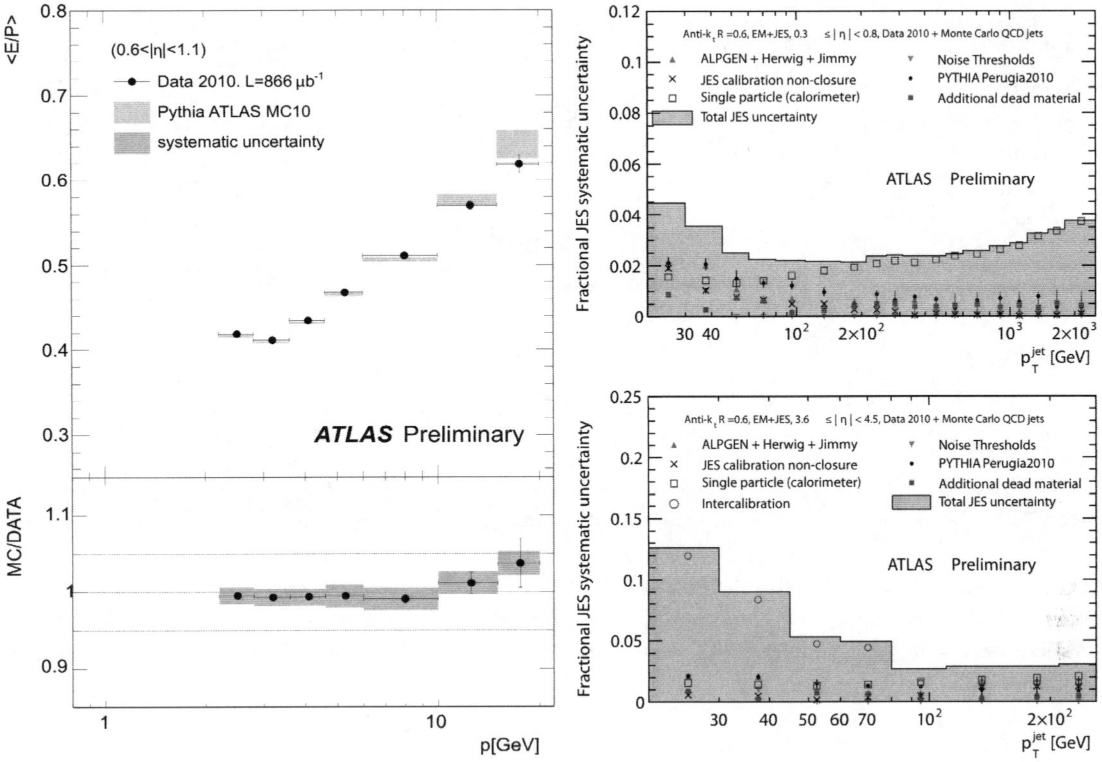

Figure 6. Single charged hadron response measured by the average calorimeter response over the track momentum $\langle E/p \rangle$, as function of p (left, Ref. [25]). The upper right figure from Ref. [23] shows the total (fractional) systematic jet energy scale uncertainty for central jets ($0.3 \leq |\eta_{\rm jet}| < 0.8$) and the contributing error sources as function of the jet $p_{\rm T}$, while the lower right figure shows the same for forward jets ($3.6 \leq |\eta_{\rm jet}| < 4.5$).

5. Systematic jet energy scale uncertainty

The systematic jet energy scale uncertainty for the present default jet calibration sequence is determined by the model dependence of the calibration functions and the precision of the data driven corrections. While the careful evaluation of the pile-up correction shows little systematic uncertainty in this procedure, the response correction receives contributions from multiple sources. These include physics modeling, detector simulation (geometry and signal modeling), uncertainties related to the actual beam spot location, and uncertainties related to observed differences between experimental and simulated response. A careful analysis of this last aspect includes the determination of the uncertainty of the charged hadron response from isolated tracks in collision events with momentum $2 < p < 20$ GeV, the uncertainty related to the response to higher momentum ($20 < p < 350$ GeV) charged hadrons from test beams, the basic electromagnetic scale uncertainty from $Z \rightarrow e^+e^-$ in pp collisions, and estimates of response uncertainties for neutral hadrons of all energies, and all hadrons with $p > 400$ GeV (all summarized in Ref. [25]). Additional contributions to the systematic jet energy scale uncertainty arise from threshold effects in the cell clustering algorithm, and the intercalibration in η with di-jet events [26], especially for regions beyond the tracking acceptance ($|\eta| < 2.5$). Figure 6 shows the excellent understanding of the single charged hadron response in ATLAS, with systematic

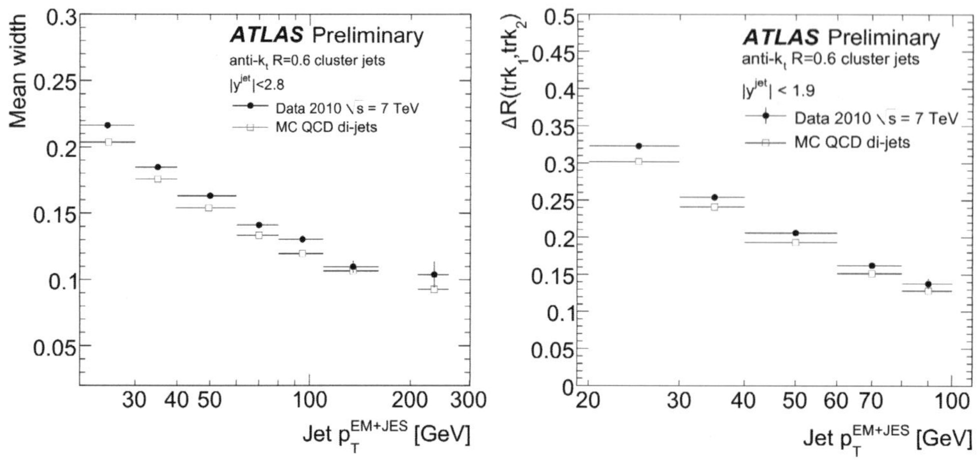

Figure 7. The average jet width $\langle w_{\rm jet}\rangle$ as function the jet $p_{\rm T}$, measured by the ATLAS calorimeters using Eq.(2), for anti-$k_{\rm T}$ jets with $R = 0.6$ and $p_{\rm T} > 20$ GeV (left). The observed disagreement with simulations based on pre-LHC collision modeling parameters for parton showering, radiation and soft QCD is confirmed by the average distance between the two hardest charged tracks measured in the ID pointing to these jet, as seen in the right figure (both figures from Ref. [16]).

uncertainties determined from the comparison with simulations of less than 2% for $p < 10$ GeV and less than 5% for $10 < p < 20$ GeV. The same figure also shows that this is the dominating uncertainty for central jets with $p_{\rm T} > 100$ GeV. The η intercalibration introduces the largest fractional errors for forward jets with $p_{\rm T} < 100$ GeV, at similar jet energies. Additional in-situ evaluations of jet energy scale uncertainties and extrapolation techniques are discussed in Ref. [2].

6. Observations from jet reconstruction

Even with the relatively small collision data sample collected by ATLAS in 2010 it is possible to demonstrate very good jet reconstruction performance. As discussed in the previous section, the systematic jet energy scale uncertainty is between about $2 - 4\%$ in the central region. The measurement of other jet features like width and shape are well understood as well. For example, the calorimeter measurement of the jet width $w_{\rm jet}$, defined as

$$w_{\rm jet} = \frac{1}{E_{\rm T,jet}} \sum_{i=1}^{N_c} r_i \times E_{{\rm T},i}, \text{ with } E_{\rm T,jet} = \sum_{i=1}^{N_c} E_{{\rm T},i} \text{ and } r_i = \sqrt{(\phi_i - \phi_{\rm jet})^2 + (\eta_i - \eta_{\rm jet})^2}, \quad (2)$$

with N_c being the number of clusters in the jet, indicates wider jets in data than in early[3] jet modeling. This is confirmed by a complementary width measurement from the ID, the distance $\Delta R({\rm trk}_1, {\rm trk}_2)$ between the two hardest tracks ${\rm trk}_1, {\rm trk}_2$ pointing to the jet, see Fig. 7 from Ref. [16].

The jet reconstruction and calibration approach presented here is appropriate for many physics analysis, see Refs. [3–5] for examples shown at this workshop. Other precise QCD measurements performed by ATLAS with the 2010 data include jet topologies and shapes, both

[3] No soft QCD tuning to actual LHC data yet in the model.

of which also provide input to tuning the soft QCD modeling parameters. Figure 8 from Ref. [27] shows the unfolded jet shape measurement in terms of the energy containment fraction $\Psi(r)$, which is the fraction of jet energy contained inside a cone of $r = 0.3$ around the jet axis, as function of the jet $p_{\rm T}$. The shape indicates a changing mixture of quark and gluon initiated jets in the final state with increasing jet $p_{\rm T}$, convoluted with perturbative QCD effects from the running strong coupling, as expected from similar studies at the Tevatron [28].

A precise measurement sensitive to QCD dynamics, in particular concerning jet production by additional parton emission, is the angular de-correlation between the two leading jets, which has first been done at the Tevatron [29] and is also available from ATLAS in Ref. [30], see Fig. 8. The increasing azimuthal de-correlation with increasing number of jets in the events is well modeled by Pythia.

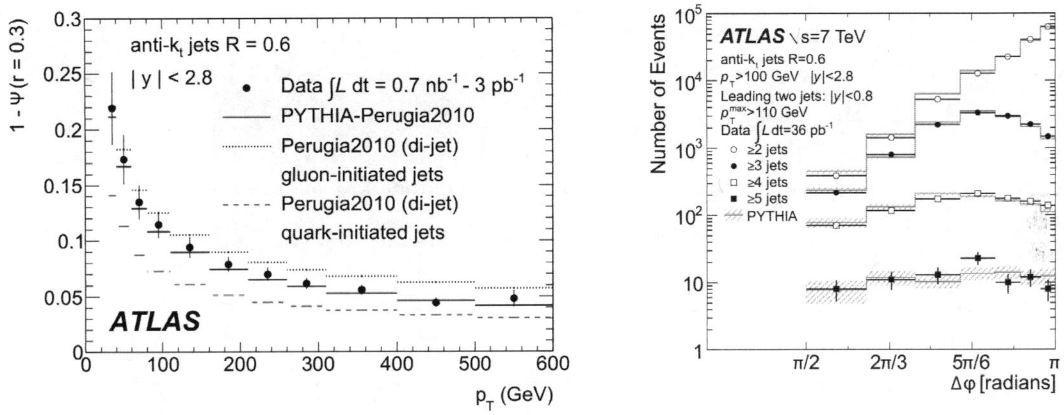

Figure 8. The fraction of jet energy outside a cone of $r = 0.3$ around the jet axis, measured in terms of the relative energy containment $\Psi(r)$, as function of the jet $p_{\rm T}$ (left, from Ref. [27]). The azimuthal distance between the two leading jets in events with $n_{\rm jets} \geq 2$ to 5 is shown in the right figure from Ref. [30], which indicates increasing de-correlation with increasing number of additional jets.

7. Conclusion and outlook

During 2010, the first year of pp collisions at $\sqrt{s} = 7$ TeV in LHC, ATLAS collected a large amount of data for studying jet properties, Standard Model jet physics, and jet topologies as search tools for new physics. A very good jet reconstruction performance, even with a simplified calibration model (the present default), has been achieved with systematic jet energy scale uncertainties as low as 2% for central jets. Early measurements of jet shapes helped improving the modeling of parton showers and radiation in the new kinematic domain now accessible. Due to the precision achieved in jet reconstruction, measurements of inclusive jet and di-jet cross sections, and angular de-correlation, among others, in this high energy domain provided important tests of the Standard Model and thus a solid base for background determinations for searches with jets in the final state. Further improvements presently under commissioning include dynamic hadronic calibration, which improves the jet energy resolution and is useful for jet substructure analysis, e.g. to understand sub-jet calibrations in jets from boosted heavy particle decays.

Acknowledgements
The results presented here are a very small subset of performance and physics results produced by the ATLAS collaboration and reflect the work of a large number of individuals contributing to the common effort of building and operating the experiment, and performing the analysis, all of whom I wish to thank warmly. In addition, I like to acknowledge the immense effort of the LHC community and CERN as a host in the construction and operation of the LHC, without which none of the results presented here would be possible. I also like to thank all funding agencies for their continuing support of the many member institutions and individuals in ATLAS. Last but not least I like to thank the organizers of this very useful workshop for the opportunity to present the jet reconstruction strategies in ATLAS.

References

[1] Aad G *et al.* (ATLAS Collaboration) 2008 *JINST* **3** S08003
[2] Schouten D 2011 In-situ measurements of the jet energy scale in ATLAS (these proceedings)
[3] Vives Vaque F 2011 Inclusive jet cross section measurement in ATLAS (these proceedings)
[4] Neusiedl A 2011 Top mass reconstruction in ATLAS (these proceedings)
[5] Salvachua Ferrando B 2011 Searches for new physics with jets in ATLAS (these proceedings)
[6] Aharrouche M *et al.* 2010 *Nucl.Instrum.Meth.* **A614** 400–432
[7] Cojocaru C *et al.* 2004 *Nucl.Instrum.Meth.* **A531** 481–514
[8] Pinfold J *et al.* 2008 *Nucl.Instrum.Meth.* **A593** 324–342
[9] Archambault J *et al.* 2008 *JINST* **3** P02002
[10] Adragna P *et al.* 2009 *Nucl.Instrum.Meth.* **A606** 362–394
[11] Aad G *et al.* (ATLAS Collaboration) 2010 *Eur.Phys.J.* **C70** 723–753
[12] Aad G *et al.* (ATLAS Collaboration) 2010 *Eur.Phys.J.* **C70** 1193–1236
[13] Aad G *et al.* (ATLAS Collaboration) 2011 (*Preprint* 1103.1816)
[14] Lampl W *et al.* 2008 ATL-LARG-PUB-2008-002
[15] Sjostrand T, Mrenna S and Skands P Z 2006 *JHEP* **05** 026
[16] ATLAS Collaboration 2010 ATLAS-CONF-2010-053
[17] Agostinelli S *et al.* (GEANT4) 2003 *Nucl. Instrum. Meth.* **A506** 250–303
[18] Aad G *et al.* (ATLAS Collaboration) 2010 *Eur. Phys. J.* **C70** 823–874
[19] M. Cacciari, G. P. Salam, G. Soyez, http://fastjet.fr/
[20] Salam G P and Soyez G 2007 *JHEP* **0705** 086
[21] Cacciari M, Salam G P and Soyez G 2008 *JHEP* **04** 063
[22] ATLAS Collaboration 2011 ATLAS-CONF-2011-030
[23] ATLAS Collaboration 2011 ATLAS-CONF-2011-032
[24] ATLAS Collaboration 2010 ATLAS-CONF-2010-056
[25] ATLAS Collaboration 2011 ATLAS-CONF-2011-028
[26] ATLAS Collaboration 2011 ATLAS-CONF-2011-014
[27] Aad G *et al.* (Atlas Collaboration) 2011 *Phys.Rev.* **D83** 052003
[28] Acosta D E *et al.* (CDF Collaboration) 2005 *Phys.Rev.* **D71** 112002
[29] Abazov V *et al.* (D0 Collaboration) 2005 *Phys.Rev.Lett.* **94** 221801
[30] Aad G *et al.* (ATLAS Collaboration) 2011 *Phys.Rev.Lett.* **106** 172002

In Situ Measurements of the Jet Energy Scale in ATLAS

Doug Schouten

TRIUMF, 4004 Wesbrook Mall Vancouver, BC, Canada, V6T 2A3

E-mail: doug.schouten@triumf.ca

Abstract. Hadron jets are the most commonly observed objects in p-p collisions at the Large Hadron Collider at CERN. Because of this, they are part of the final state of almost any process, and are an important probe in searches for extensions of the Standard Model.

A precise knowledge of the energy calibration for jets is difficult to ascertain for a number of reasons, and is a necessary ingredient in the ATLAS experimental program. This report presents in situ techniques and results for the jet energy scale at ATLAS using recent collision data. This report demonstrates an understanding of the necessary jet energy corrections to within $\approx 4\%$ in the central region of the calorimeter.

1. Introduction

Over the course of 2010, the ATLAS detector [1] collected data from proton-proton collisions delivered by the Large Hadron Collider (LHC) at a center-of-mass energy of $\sqrt{s} = 7$ TeV. Understanding and measuring the performance of jets in these data is crucial for many measurements with ATLAS. Indeed, the uncertainty of the jet energy scale (JES) is the dominant experimental uncertainty for numerous physics results, for example the cross-section measurement of the inclusive jets, dijets [2] and multijets, vector boson accompanied by jets [3], and for new physics searches with jets in the final state as in [4], to name a few.

An estimate of the JES uncertainty has been derived in Ref. [5] using in situ measurements of the single hadron response [6]. This report summarizes the JES systematic uncertainty based on analysis of the first year of ATLAS data, and results from the JES uncertainty validation using in situ techniques.

1.1. *The ATLAS Detector*

The ATLAS detector consists of a tracking system immersed in a 2 T solenoidal magnetic field up to a pseudorapidity[1] $|\eta| < 2.5$, sampling electromagnetic and hadronic calorimeters up to $|\eta| < 4.9$, and muon chambers in a toroidal magnetic field. A more detailed description of the ATLAS experiment can be found elsewhere [1]. Jets are reconstructed using the ATLAS calorimeters. The electromagnetic calorimeters use liquid argon as the active material and lead for the absorbers, and they cover the fiducial region up to $|\eta| < 3.2$. Liquid argon presamplers

[1] The ATLAS Coordinate System is a right-handed system with the x-axis pointing to the centre of the LHC ring, the y-axis pointing upwards and the z-axis following the beam direction. The pseudorapidity η is an approximation for rapidity y in the high energy (massless) limit, and is related to the polar angle θ by $\eta = -\ln(\tan \frac{\theta}{2})$.

situated in front of the electromagnetic calorimeter allow for corrections of upstream energy loss due to the presence of the barrel solenoid and inner detector cryostat. The hadronic calorimeters employ plastic scintillator and steel for the barrel and extended barrels (covering $0 < |\eta| < 0.8$ and $0.8 < |\eta| < 1.7$, respectively), and liquid argon and copper for the endcaps ($1.5 < |\eta| < 3.2$). The forward calorimeter is a liquid argon and tungsten/copper detector, and extends the calorimetry up to $|\eta| < 4.9$ for nearly hermetic coverage.

1.2. Jet Reconstruction

Jets are reconstructed by the FastJet software [7], using the anti-k_t algorithm [8] with distance parameters $R = 0.6$ and $R = 0.4$. In the following, jets with distance parameter $R = 0.6$ are discussed. The results for jets with $R = 0.4$ are similar and are excluded for brevity. The constituents of calorimeter jets are topological clusters (topo-clusters) [9] that group together calorimeter cells and are designed to follow the shower development. The topo-cluster algorithm starts from a seed cell, whose signal-to-noise ratio is above a threshold of 4. Cells that are topologically connected to the seed and that have a signal-to-noise ratio of at least 2 are included iteratively, and finally all directly neighboring cells are added to the topo-cluster. A topo-cluster is defined to have an energy equal to the energy sum of all the included cells, zero mass and a reconstructed direction along the unit vector originating from the center of the detector, pointing to the energy-weighted topo-cluster barycenter. Note that the ATLAS calorimetry is designed with a pointing geometry that projects to the center of the coordinate system.

2. Calibration of the Jet Energy Scale

The goal of the JES calibration is to correct the E and $\vec{p}$ of jets measured in the calorimeter, to the corresponding particle reference jets (which are jets built from all final state particles in an event, except muons and neutrinos). This calibration must account for a number of detector effects:

- calorimeter response non-compensation ($e/h > 1.3$ in ATLAS);
- inactive regions (sometimes known as dead material) and punch through, in which the calorimeter does not fully contain the shower;
- calorimeter signal definition inefficiencies due to clustering noise thresholds, or jet width parameter;
- additional energy deposited in the reconstructed jet from pileup.

For the 2010 analyses, ATLAS has employed a technique, known as `EMJES`, which is based on calibration factors derived from Monte Carlo simulations. The steps involved in this scheme are shown in Figure 1. The calibration begins at the calorimeter electromagnetic scale, which is validated with $Z^0 \to e^+e^-$ events and minimum-ionizing muons. Then, a correction to account for the additional energy contribution due to pileup is applied. This correction is derived from minimum bias data as a function of the number of reconstructed primary vertices in the event and the jet η. It takes into account the average additional energy deposited in a fixed grid of 0.1×0.1 in the (η, ϕ)-plane (called towers) and the average number of such towers in a jet. The average energy deposition per tower and average tower multiplicity per jet are shown in Figure 2.

Because the jet reconstruction is performed using topoclusters assuming an interaction point at $(x, y, z) = (0, 0, 0)$, an additional correction is needed to adjust the momentum of the jet such that the momentum unit vector points from the leading primary vertex in the event[2] from which the jet originates.

[2] The leading primary vertex is defined here as the reconstructed vertex with the highest $\sum_{i \in \{\text{tracks}\}} \left(p_T^{(i)}\right)^2$.

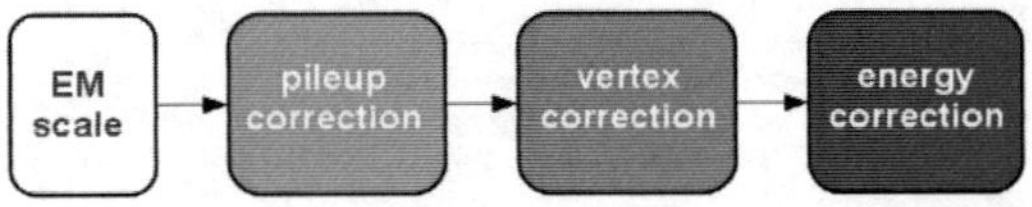

Figure 1. The various steps involved in the EMJES calibration scheme: the raw electromagnetic scale, a correction for pileup, a correction to the primary vertex from the detector centroid $(0, 0, 0)$, and finally a correction for the calorimeter response.

The final step of the EMJES jet calibration restores the reconstructed jet energy to the energy of the reference jet in the Monte Carlo simulations. Since pile-up effects have already been corrected for, the simulation samples used to derive the calibration do not include multiple proton-proton interactions within the same bunch crossing. The correction is defined in a fine η segmentation and is applied as a simple scaling factor $C = C(E, \eta)$, where E and η are the energy and pseudo-rapidity of the jet with the first two corrections applied. This correction is shown in Figure 3.

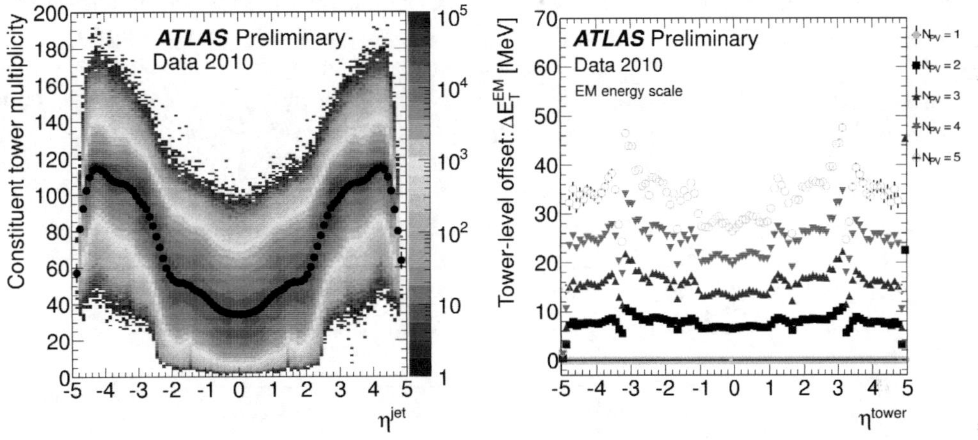

Figure 2. The average tower multiplicity in each jet as a function of pseudo-rapidity (left) and the average tower offset E_T in minimum bias events, shown for various numbers of reconstructed primary vertices (i.e., different levels of pileup), and as a function of the η centroid of the tower (right).

3. Systematic Uncertainties

The overall strategy used to ascertain the validity of the calibration procedure [10], is to evaluate the jet energy scale by roughly factorizing the components of EMJES, verifying that the Monte Carlo description of each component in the data is correct. In other words, the role of the in situ measurements is to provide systematic uncertainties on the jet energy scale. Contributions to the overall JES systematic uncertainty arising from non-closure of the calibration in the simulations, jet fragmentation uncertainty, effects of the noise thresholds used in clustering, or from mismodeling of the inner detector in the detector simulation, are derived from dedicated simulation studies. The largest components of the uncertainty, arising from description of the calorimeter response, are derived with two data-driven techniques: single particle response from E/p and test-beam measurements in the central region ($|\eta| < 0.8$), and relative calibration with QCD dijet events for the endcap and forward regions ($|\eta| > 0.8$).

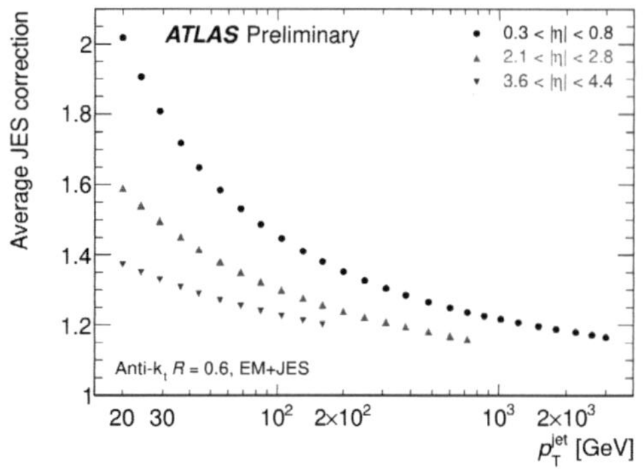

Figure 3. The energy correction factor C described in the text, shown as a function of jet p_T for various η ranges.

3.1. Single Particle Response

The core idea is to measure the response for isolated single particles by comparing the energy (calorimeter) and momentum (tracking) measurements, namely E/p. Then, pseudo experiments in Monte Carlo simulations are employed to extrapolate the single particle response uncertainty to a jet response uncertainty. Although the translation from single particles to a jet context is non-trivial, it has been exhaustively cross-checked and found to have small uncertainty.

For charged particles in the momentum range $0.5 < p < 20$ GeV the E/p measured in situ is used to determine the response [6]. In order to account for the neutral background component to the measured E/p, the background is estimated by looking in an annulus $0.1 < \Delta R = \sqrt{\Delta\eta^2 + \Delta\phi^2} < 0.2$ around minimum ionizing particles in the EM calorimeter. This contribution is then subtracted from the measured E/p ratio. A comparison of E/p in data to that in Monte Carlo is shown in Figure 4.

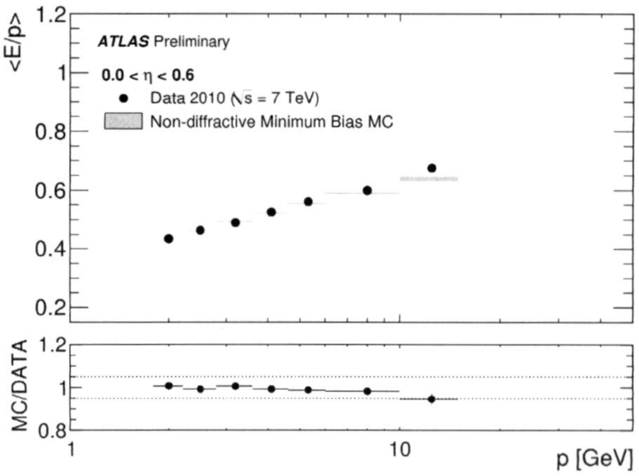

Figure 4. The E/p measured in data (black points) compared to the same measurement in Monte Carlo simulations (shaded green histogram), as a function of particle momentum. The ratio of data to Monte Carlo is also shown, in which the dashed lines are located at $\pm 5\%$ of unity.

The response for charged particles in the momenutm range $20 < p < 400$ GeV is estimated from test-beam measurements using a nearly exact replica of a slice of the ATLAS detector. An uncertainty that accounts for the differences in the test-beam setup compared to the ATLAS detector "as built" is included in this measurement.

Additional uncertainties that are included in the single particle extrapolation account for effects related to the calorimeter acceptance, uncertainties related to particles with $p > 400$ GeV, and uncertainties for the response of neutral hadrons.

3.2. QCD Dijet Relative Response

For jets in the region $|\eta| > 0.8$, the response for the central region is extrapolated using a dijet balance technique [11]. This procedure derives a calibration for a jet under the assumption of momentum balance of the dijet system at the particle (reference) level. Supposing that in a QCD dijet event one of the jets lands in η bin i and the other in bin j, one can define an event asymmetry

$$A_{ij} = \frac{p_T^i - p_T^j}{\frac{1}{2}\left(p_T^i + p_T^j\right)} \tag{1}$$

so that $R_{ij} = \frac{2-\langle A_{ij}\rangle}{2+\langle A_{ij}\rangle}$ is the average ratio of the relative response coefficients, α_i/α_j, over many events. After selecting events with a QCD dijet topology, one can calculate a matrix R_{ij} and minimize

$$S = \sum_{i<j}\left(\frac{1}{\Delta R_{ij}}(\alpha_j R_{ij} - \alpha_i)\right)^2 + \chi(\{\alpha_i\}), \tag{2}$$

where χ is a constraint on the central response corrections (which are set to unity, since the central response is fixed) and ΔR_{ij} is the matrix of statistical uncertainties for R_{ij}. Minimizing S results in coefficients $\{\alpha_i\}$ that equalize the calorimeter response in all η bins to the central region. This procedure is applied in bins of transverse momenta to derive corrections $\{\alpha_i(p_T)\}$.

For the relative response measurement a combination of minimum bias and jet triggers are chosen to ensure that only events on the plateau of each trigger are used in the analysis, and events are then selected that contain two jets well-separated in ϕ, and with no high p_T third jet.

An inconsistency is noticed in the relative response coefficients calculated with different Monte Carlo generators. Since there is no a priori reason to select one generator over another, the RMS deviation between Monte Carlo and data is taken as the systematic uncertainty in the dijet balance. Then the total uncertainty in the endcap region is given as

$$\sigma_{|\eta|>0.8} = \sigma_{|\eta|<0.8} \oplus \sigma_{\text{dijet}}.$$

The systematic uncertainty from dijet balance is shown in Figure 5.

4. In Situ Validations

In order to validate the overall jet energy scale uncertainty, various measurements are performed that are directly sensitive to the JES. Comparisons of these measurements between Monte Carlo and data provide validation that the JES uncertainty estimate is correct.

4.1. Photon + Jet Results

Because photons are well-measured objects, one can directly measure the jet response by using the principle of momentum balance between a photon and recoil jet in photon + jet events [12]. Two techniques are used: the missing E_T projection fraction (MPF) method and the direct balance method. The two are in some sense complementary: the MPF method measures the total calorimeter response, whereas the direct balance method is also sensitive to the leakage of energy outside of the calorimeter jet. The ratio of data to Monte Carlo for the MPF and direct balance measurements are shown in Figure 6, along with the associated systematic and statistical uncertainties. The Monte Carlo and data agree within 5% for $25 < p_T < 250$ GeV in the central region.

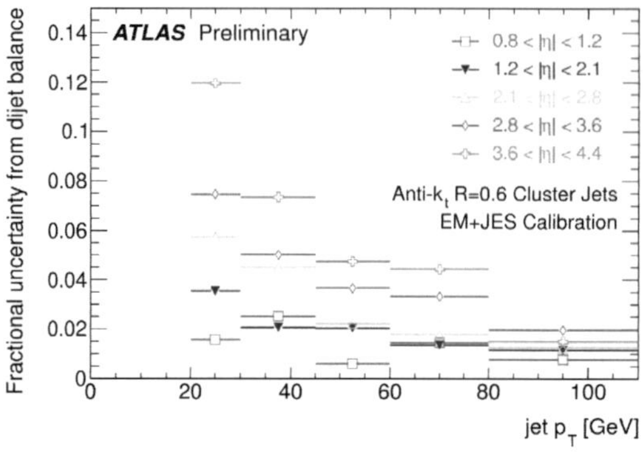

Figure 5. Systematic uncertainty derived for the relative correction using dijet balance as a function of jet p_T.

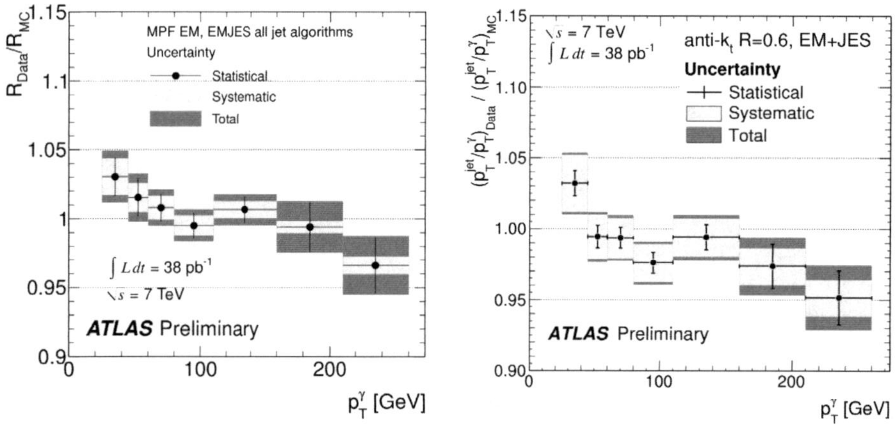

Figure 6. The ratio of data to Monte Carlo for the MPF (left) and direct balance (right) as a function of photon p_T in photon + jet events. The yellow band indicates the size of the systematic uncertainty due to dijet background, photon energy scale and gluon radiation modeling.

4.2. QCD Multijet Balance

If jets at low transverse momentum are well calibrated, one can assess the calibration of jets at high p_T by balancing against a recoil system of at least two lower p_T jets, both of which are in a region for which a calibration is known [13]. This method allows probing the jet energy scale up to the TeV regime. A multi-jet topology is selected by demanding that events contain at least three jets and that the p_T of the recoil system is evenly distributed. The recoil system and leading jet are also required to be well-separated in ϕ. Considering effects due to calibration of low p_T jets, gluon radiation, and the presence of nearby jets on the response, the total systematic effects are evaluated to be 4%. The ratio of the multi-jet balance in data to Monte Carlo for the region $0 < |\eta| < 2.8$ is shown in Figure 7.

5. Summary

The jet energy scale uncertainty for jets reconstructed with the anti-k_t algorithm with distance parameters $R = 0.4$ and $R = 0.6$, has been estimated in ATLAS. The JES and its uncertainty are evaluated up to the kinematic limit (jet energies up to 3.5 TeV) and for pseudorapidities $|\eta| < 4.5$. Uncertainties are derived using primarily data-driven approaches. The results are

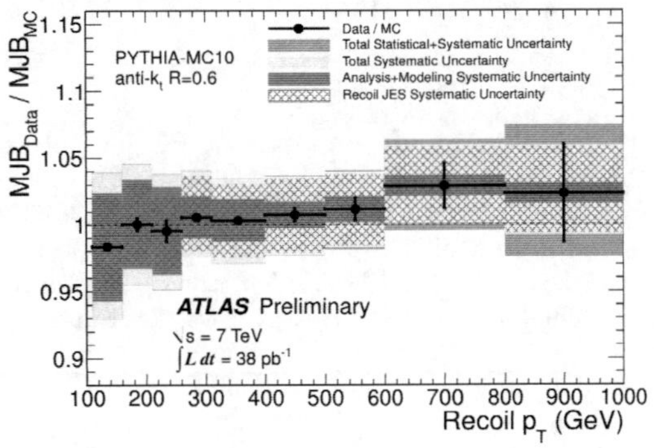

Figure 7. The ratio of the multi-jet balance in data to Monte Carlo as a function of the p_T of the recoil system. The data and Monte Carlo agree within systematic and statistical uncertainties.

further verified using photon + jet and multijet balancing methods in situ. A summary of all the validation methods and the JES uncertainty is shown in Figure 8.

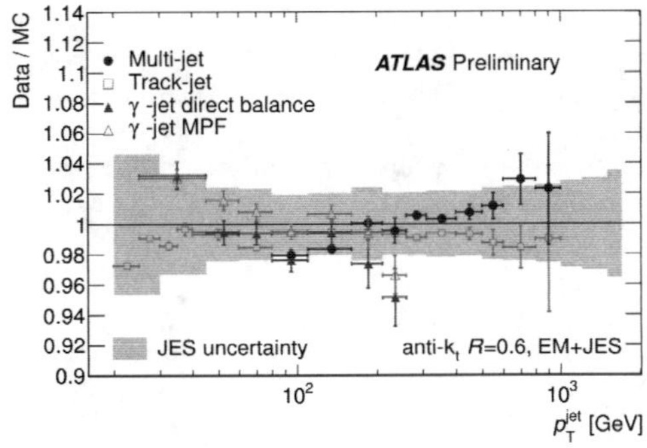

Figure 8. The jet energy scale uncertainty (blue band) in the central region compared to data and Monte Carlo ratios calculated in various validation measurements. The JES uncertainty agrees with the measurements over a large range of transverse momenta.

References

[1] ATLAS Collaboration 2008 The ATLAS Experiment at the CERN Large Hadron Collider *JINST* **3** S08003.

[2] ATLAS Collaboration 2011 Measurement of inclusive jet and dijet cross sections in proton-proton collisions at 7 TeV centre-of-mass energy with the ATLAS detector *Eur. Phys. J. C* **71** 1512 (*Preprint* arXiv:1009.5908).

[3] ATLAS Collaboration 2010 Measurement of the production cross section for W-bosons in association with jets in pp collisions at $\sqrt{s} = 7$ TeV with the ATLAS detector *Phys. Lett. B* (*Preprint* arXiv:1012.5382).

[4] ATLAS Collaboration 2010 Search for New Particles in Two-Jet Final States in 7 TeV Proton-Proton Collisions with the ATLAS Detector at the LHC *Phys. Rev. Lett.* **105** 161801 (*Preprint* arXiv:1008.2461).

[5] ATLAS Collaboration 2011 Update on the jet energy scale systematic uncertainty for jets produced in proton-proton collisions at $\sqrt{s} = 7$ TeV measured with the ATLAS detector *ATLAS-CONF-2011-007*.

[6] ATLAS Collaboration 2010 ATLAS calorimeter response to single isolated hadrons and estimation of the calorimeter jet scale uncertainty *ATLAS-CONF-2010-052*.

[7] M. Cacciari and G. P. Salam 2006 Dispelling the N^3 myth for the k_t jet-finder *Phys. Lett. B* **641** no. 1, 57 - 61.

[8] M. Cacciari, G. P. Salam, and G. Soyez 2008 The anti-kt jet clustering algorithm *JHEP* **04** 063 (*Preprint* arXiv:0802.1189).

[9] W. Lampl et al. 2008 Calorimeter clustering algorithms: description and performance *ATL-LARG-PUB-2008-002*.

[10] ATLAS Collaboration 2011 Jet energy scale and its systematic uncertainty in proton-proton collisions at $\sqrt{s}$ = 7 TeV with ATLAS 2010 data *ATLAS-CONF-2011-032*.

[11] ATLAS Collaboration 2011 In-situ pseudorapidity intercalibration for evaluation of jet energy scale uncertainty using dijet events in proton-proton collisions at $\sqrt{s} = 7$ TeV *ATLAS-CONF-2011-014*.

[12] ATLAS Collaboration 2011 Determination of the jet energy measurement uncertainty using photon-jet events in proton-proton collisions at $\sqrt{s} = 7$ TeV *ATLAS-CONF-2011-031*.

[13] ATLAS Collaboration 2010 Measurements of multijet production cross sections in proton-proton collisions at 7 TeV center-of-mass energy with the ATLAS Detector *ATLAS-CONF-2010-084*.

Measurement of the Jet Energy Scale at DØ

Gianluca Petrillo *for the DØ collaboration*

University of Rochester, 500 Wilson Boulevard, Rochester, NY 14627-0171, USA

E-mail: `petrillo@fnal.gov`

Abstract. Jets of hadrons are the most commonly produced objects at high-energy hadronic colliders. It is therefore important to identify and quantify jet energies in events. However, collisions involving interesting physical processes can overlap in time with additional collisions. This is especially troublesome at high luminosities, as such additional energy depositions in detectors can compromise the reconstruction of jets. The DØ detector at Fermilab relies mainly on the energy information obtained using its uranium/liquid-argon (LAr) sampling calorimetry. The experiment has developed a set of corrections, collectively known as the jet energy scale (JES), that translate jet energies measured in the detector to the energies carried by the produced particles. This correction is extracted from two-body events, separately for those observed in the detector and for the ones from Monte-Carlo simulation. The corrections include effects from the jet reconstruction algorithm, the response of the calorimeter and the impact of simultaneous interactions. We describe the JES correction used at DØ, focusing on the main contributions and discuss the precision achieved in the measurement of jet energy.

1. Instrumentation

The Tevatron at Fermi National Accelerator Laboratory accelerates protons and antiprotons that collide at a centre of mass energy of 1.96 TeV. Before the shutdown planned for October 2011, the accelerator is expected to deliver to the experiments about $12\,\mathrm{fb}^{-1}$, which will provide CDF and DØ experiments about $10\ \mathrm{fb}^{-1}$ of good data. This achievement is a result of the outstanding performance of the accelerator, including an instantaneous luminosity ($\mathcal{L}$) that has greatly exceeded original design specifications. In fact, the initial $\mathcal{L}$ of each store is often $\approx 3 \cdot 10^{32}\ \mathrm{cm}^{-2}\mathrm{s}^{-1}$. DØ data has been taken with an average of the order of $10^{32}\ \mathrm{cm}^{-2}\mathrm{s}^{-1}$. This, together with a frequency of almost 1.7 million bunch crossings per second at a cross section of 60 mb, causes on average 4 interactions for each bunch crossing.

The DØ detector [1] is a multi-purpose detector made of an inner tracking section; a calorimetry section; and outer tracking detectors for identifying muons and measuring their momenta. Most of the detector is used in the reconstruction of jets: the reconstruction of the primary interaction vertex, based on tracks, provides a reference point to reconstruct the jet, any associated muon is used to improve the estimation of jet energy, but the main information is extracted from the calorimeter.

The core of DØ calorimetry is a sampling calorimeter (Fig. 1) that uses liquid argon as active material and depleted uranium (or copper and steel in some forward regions) as absorber [2]. To keep the argon in liquid state, the whole calorimeter is contained in cryogenic tanks. The detector is split into one *central calorimeter* (CC) and two *end-cap* calorimeters (EC), each in separate cryostats. The region between the cryostats, called *intercryostat region* (ICR), is instrumented with scintillators to improve the coverage of the calorimeter. The calorimeter is fully sensitive in azimuthal angle (φ_d) and up to 4.1 units in pseudorapidity ($|\eta_\mathrm{d}|$). The effect of the gap in the ICR is most evident for jets with axis on $0.9 < |\eta_\mathrm{d}| < 1.6$. Central and forward preshower scintillator elements complete the calorimeter.

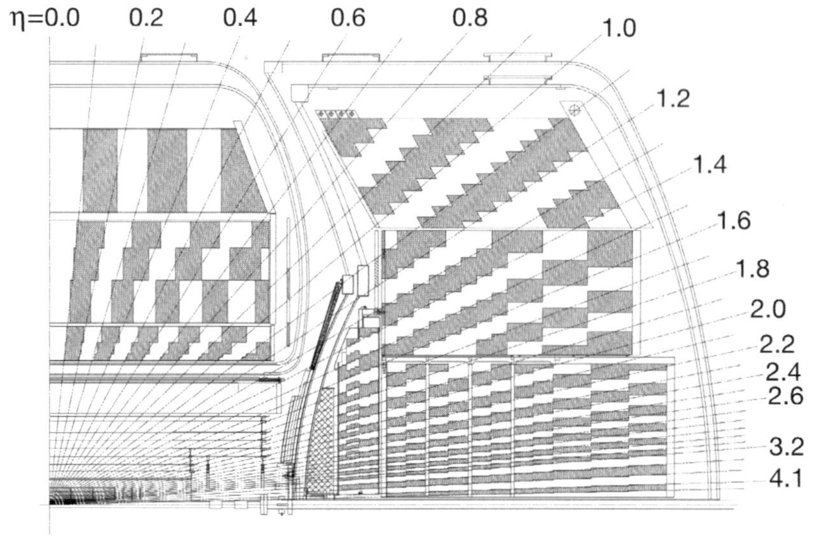

Figure 1. Section of the DØ calorimeter. Projective calorimeter towers are marked in alternating shadings of gray and white.

The overall thickness of the electromagnetic layers of the detector can be measured in 20 radiation lengths (X_0) and the entire calorimeter varies from 7.2 to 8.0 strong interaction lengths (λ_I), while the material in front of the calorimeter is about 4 X_0 thick.

The calorimeter is segmented longitudinally, and split into 17 layers grouped into electromagnetic parts, three fine hadronic parts layers and a coarse-hadronic part. Cells are aligned across layers in a roughly pointing-geometry to form "calorimeter towers". The cells have a size of $\eta_\text{d} \times \varphi_\text{d} = 0.1 \times 0.1$.

2. Algorithms

2.1. Zero suppression

There are about 50000 cells in the calorimeter, each read out independently. To reduce the bandwidth needed to store this amount of data at a 1.7 MHz event rate, a filter is applied, commonly known as "zero suppression" (ZS).

The ZS algorithm discards the readings from the cells that have no relevant energy deposition.
Each cell has a noise level (σ) and a pedestal setting, which correspond respectively to the RMS and average of the cell readings without collisions. The readout electronics discards the readings below 1.5σ (Fig. 2.) The others are analysed by an algorithm (called "T42") which, using the geometry of the detector, discards all cells with readings below 2.5σ, and the ones with reading below 4σ that do not have any neighbouring cell with reading above 4σ (Fig. 3.)

This procedure causes an underestimation of the measured energy, which is corrected though the jet energy scale.

2.2. Jet reconstruction

DØ uses a fixed-cone algorithm with midpoints [3] to reconstruct jets from energy depositions in calorimeter cells.
The algorithm starts from *seeds* consisting of highly energetic calorimeter towers; a calorimeter tower is a group of cells aligned on a given direction from the centre of the detector, spanning all the longitudinal

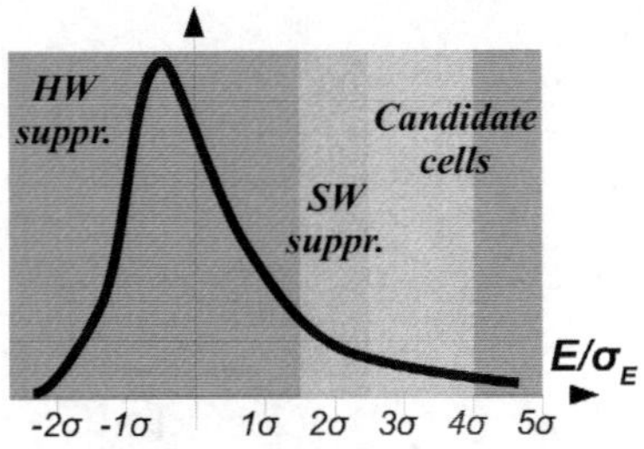

Figure 2. Hardware zero suppression: cutoff on the noise, with the curve representing a typical distribution in units of RMS cell noise σ, with the origin set to the cell mean pedestal value (grey areas are always suppressed).

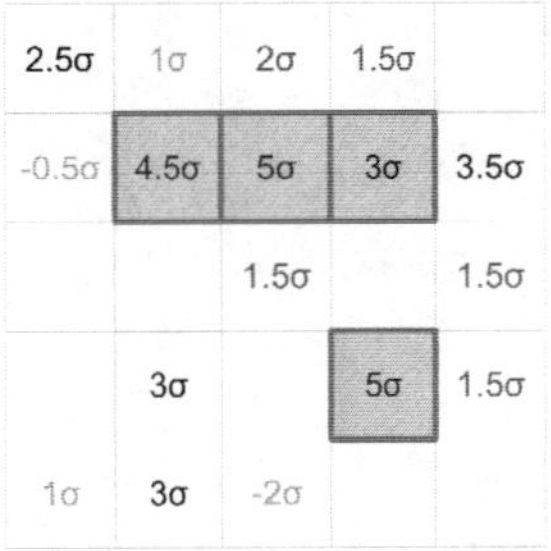

Figure 3. Software zero suppression (T42 algorithm): in this example each cell from the same layer contains its readout in σ units and the only not suppressed cells are the four cells highlighted (note that the DØ algorithm considers all the layers in the decision).

layers of the calorimeter. For the reconstruction to be stable and less sensitive to gluon radiation ("infrared safety"), additional artificial seeds, *mid points*, are added between the existing seeds.

Proto-jets are formed by drawing cones of fixed opening $\Delta R = \sqrt{(\Delta\eta)^2 + (\Delta\varphi)^2}$. The procedure is iterative, and on the first iteration the direction of the seed is used to define the axis of the cone; the jet axis is then recomputed at each iteration, and the new value is used as the axis of the cone for the next iteration, until the axis stabilises.

The "E-scheme" [3] is used to compute the jet energy and axis. The cells are treated as independent, massless objects, and the jet four-momentum is computed from $\left(E_{\text{jet}}, \vec{p}_{\text{jet}}\right) = \sum_{c \in \text{cells}} \left(E_c; E_c \hat{u}_c\right)$, where E_c and $\hat{u}_c$ are, respectively, the calibrated energy and the direction of the cell relative to the primary vertex.

If any proto-jets overlap, they are merged into one jet in case the shared energy is large, or they are split and each cell is assigned to the closest jet in $DeltaR$.

DØ uses $\Delta R = 0.5$ for most of the analyses, while $\Delta R = 0.7$ is used most often for dijet and trijet analyses, which have lower final-state multiplicity and can thereby collect more of the energy of each jet. Jet Energy Scale corrections are extracted independently for each cone size.

3. Jet Energy Scale

The evolution of the gluons and quarks ("partons") produced in the hard-scattering QCD process to jets of particles is usually described in three phases or "levels":

parton level reflects the hard scattering of two partons, which produces quarks and gluons, the seeds for jets

particle level reflects the strong potential between quarks and gluons that increases as they separate in space, creating thereby more partons that eventually bind into hadrons, with their direction clustered around their original partons. The details of the development during this phase is model dependent, but at the end of the development, short-lived hadrons decay and only relatively long-lived particles remain; these final particles (including e.g. photons, pions, nucleons, Λ^0, K, electrons, muons) enter the detector

reconstruction level reflects the interaction of the particles in the detectors as they deposit their energy through electromagnetic or strong interactions in the detector material. Some energy is "lost" as

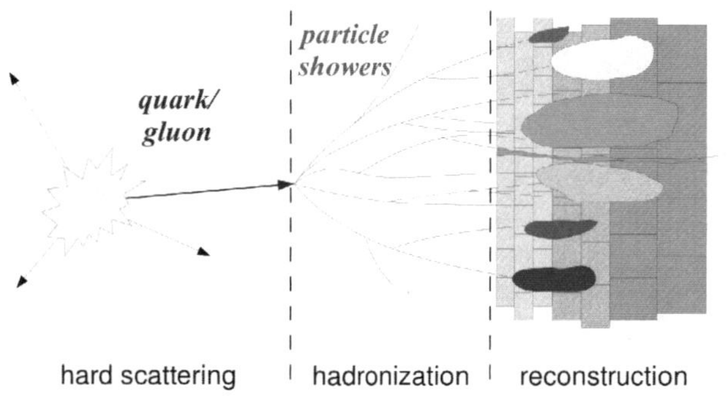

Figure 4. Development of a jet from the hard scattering.

neutrinos escape detection and nucleons are stopped in the material

We measure the energy of all objects at the reconstruction level. While our physics analyses are most concerned with the parton energies, *the goal of the Jet Energy Scale correction is to convert the reconstructed energy back to energies of particles*, which are observable and directly measurable. The next corrective step to the parton level includes other refinements such as unfolding detector and hadronization effects, which are strongly model-dependent: these are not included in the JES.

3.1. Overview of the JES correction

The Jet Energy Scale correction is summarized by this formula:

$$E_{\text{jet}}^{\text{ptcl}} = \frac{E_{\text{jet}}^{\text{meas}} - \hat{E}_{\text{O}}}{\hat{R}_{\text{jet}}\,\hat{S}_{\text{jet}}} \cdot \frac{k_{\text{O}}}{k_{\text{R}}} \tag{1}$$

The jet energy at the particle level ($E_{\text{jet}}^{\text{ptcl}}$), is computed from the detected energy $E_{\text{jet}}^{\text{meas}}$ (sum of energies from calorimeter cells), and:

- Subtracting the "offset" energy (E_{O}) from activity not associated with the main hard $p\bar{p}$ collision
- Correcting for calorimeter response (R_{jet}) to the jet of particles
- Accounting for the fact that only a fraction (S_{jet}) of the jet energy is captured in the jet cone

An estimator is used for each term, which can be biased, and therefore requires additional correction factors for e.g. energy response and offset (k_{O} and k_{R} respectively, accounting for zero-suppression effects, selection biases and contamination of the residual γ+jet events sample).
The following sections describe each one of the three main components of the correction factors.

3.2. Energy offset

The energy-offset correction accounts for energy deposited in the calorimeter which does not come from the main $p\bar{p}$ hard scattering. It includes (i) noise in the detector, both from the electronics and uranium (which, although depleted, has substantial residual activity), (ii) effects from the slow response of the calorimeter electronics relative to the bunch-crossing time ("pile up"), where the charge collected from a cell in the current crossing overlaps the remnant charge distribution from the previous collisions, (iii) additional energy from other $p\bar{p}$ interactions ("multiple interactions," MI), usually elastic, diffractive, or soft inelastic collisions. The offset correction does not account for the energy from the remnants of the proton and antiproton spectator partons that do not participate in the high-Q^2 hard scattering (the "underlying event").

The correction is defined in two steps. In the first step, the energy contained in a ring of the detector, i.e. all the cells at a fixed rapidity (all azimuthal angles and all layers), are defined as:

$$\hat{E}_{\text{O}}^{\text{ring}}(\eta_{\text{d}}, n_{\text{PV}}, \mathcal{L}) = \hat{E}_{\text{NP}}^{\text{ring}}(\eta_{\text{d}}, \mathcal{L}) + \hat{E}_{\text{MI}}^{\text{ring}}(\eta_{\text{d}}, n_{\text{PV}}, \mathcal{L}) \quad (2)$$

The energy is contributed from two sources: (i) the noise and pile-up energy ($\hat{E}_{\text{NP}}^{\text{ring}}$) depends on the rapidity of the ring, as the background activity increases in the large $|\eta_{\text{d}}|$ regions, and (ii) the instantaneous luminosity ($\mathcal{L}$). $\hat{E}_{\text{NP}}^{\text{ring}}$ is extracted from data collected using a "zero bias" (ZB) trigger, which reads out an event at random times, when a bunch crossing is expected, requiring in addition zero reconstructed interaction vertices. This ensures that the measured activity is not from in-time physical events.

The energy from multiple interactions ($\hat{E}_{\text{MI}}^{\text{ring}}$) also depends on the number of reconstructed interaction vertices (n_{PV}), which describes precisely how many interactions take place during each event. This information is extracted from data taken with a "minimum bias" (MB) trigger, which detects in-time activity in both forward ends of the detector, ensuring the presence of *some kind* of activity in the detector.

Each jet is assigned a correction depending on the area of the detector it spans, getting a contribution from each ring $\hat{E}_{\text{O}}^{\text{ring}}$ proportional to its area within that ring. Since we do not wish to subtract the energy of the main hard scatter, for each jet in an event with n_{PV} primary vertices, we use the correction for $n_{\text{PV}} - 1$ vertices (jets in events with just one primary vertex get offset contributions only from the noise and pile-up term).

The measured offset energy can become huge in extreme conditions (many primary vertices, high luminosity, very large $|\eta_{\text{d}}|$ regions), but it is usually below 25 GeV for most of the jets used in the physics analyses ($\Delta R = 0.5$ jets within $|\eta_{\text{d}}| \leq 2.5$), and only a few GeV for jets at most central η_{d} values.

The systematic uncertainties on this correction are very small and dominated by the uncertainty on the bias correction.

3.3. Energy response

The energy response is defined as the fraction of the jet energy that we measure, relative to the true energy of the particles within the jet cone. It includes: (i) effects from the difference in energy response for different particles, e.g. hadrons vs. electrons (since the calorimeter is not entirely compensating), (ii) energy not detected because converted in nucleon mass, (iii) energy loss due to inactive material (cracks) or uninstrumented regions (inter-cryostat gaps), (iv) non-uniformity of response in different regions of the calorimeter, (v) energy leaking beyond the calorimeter ("punch through") and (vi) non-linearity due to zero-suppression. In principle, it also includes undetected energy from neutrinos and reduced response from muons; a more precise evaluation of energy from muons is added as an additional correction including information from the muon detectors.

To evaluate the calorimeter response, a "tag and probe" method is used in studies of two-body events of simplest topology:

γ+jet is the most important process, defining a relation between jet energy and the energy of an electromagnetic object, the photon, assumed to be well measured;

dijet events are used to enhance statistics for the estimation of the uniformity of response, once the absolute energy scale for some jets is set through γ+jet sample. In a second step, these jets can be used as tag objects for dijet events;

$Z(\to e^+e^-)$+jet events, although excellent candidates for such studies, have not been used because of low statistics of the data samples.

The above events are required to have a low number of reconstructed primary vertices, and the two objects are required to be back to back ($\Delta\varphi \geq 3$ rad), to ensure the cleanliness of the events. Also, the requirements on the tag objects are kept stringent: photon-identification criteria for the γ+jet events are very tight, and tag jets in the dijet sample are required to have a relevant fraction of their energy from charged particles pointing back to the main primary vertex.

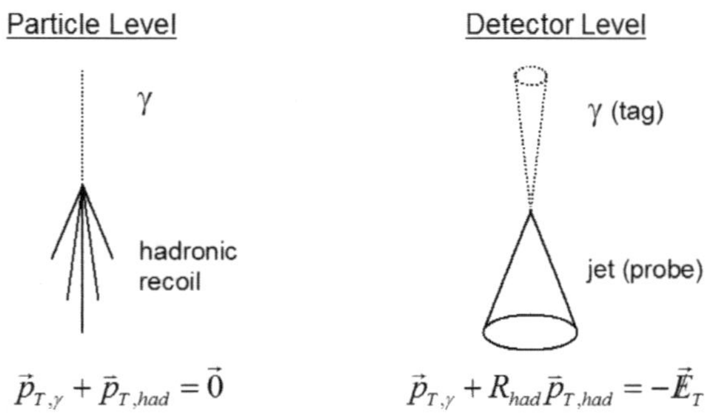

Figure 5. The Projected Missing E_T Fraction (MPF) method, used for the estimation of the response to the hadronic recoil to a photon (R_{had}), also approximated to the response to the jet (R_{jet}).

All our corrections are not parametrized as function of the jet energy, which is subject to greater uncertainties and resolution effects, and we use instead the quantity:

$$E' \equiv p_T^{\text{tag}} \cosh \eta^{\text{probe}} \tag{3}$$

where the transverse momentum p_T^{tag} of the tag is fully corrected, and we assume the direction (and the pseudorapidity η) of the probe jet from the interaction vertex to be well measured and its transverse momentum well approximated by the p_T of the tag object.

The determination of the response is based on the Projected Missing E_T Fraction (MPF) (Fig. 5), which can be described as follows. Relying on the two objects being back-to-back ($\vec{p}_{T\gamma} + \vec{p}_{T\text{had}} = 0$), the imbalance in transverse momentum $\vec{E}_T$ is assumed to reflect the deficiency in the measured jet energy, and we obtain the response as

$$R_{\text{had}} = 1 + \frac{\vec{E}_T \cdot \vec{p}_{T\gamma}}{|\vec{p}_{T\gamma}|^2} \tag{4}$$

The missing energy is computed directly from the calorimeter cells. In fact in this formula we do not consider the jet at all; instead we are measuring the response R_{had} of the calorimeter to the whole event recoiling against the tag object. A next approximation considers the response to the full recoil to be the same as the response to the jet, $R_{\text{jet}} \approx R_{\text{had}}$. This approximation is no better than $\approx 1\%$, and a bias correction is included for that. The MPF response is independent of the algorithm or cone size used to reconstruct the jet.

The first step in obtaining the response correction is the measurement of absolute response, which uses very clean γ+jet events to measure the response of very central jets ($|\eta_{\text{d}}| \leq 0.4$) relative to the well measured photon energy. The residual dijet contamination is corrected by measuring the purity of the sample, and the response from the contaminating dijet events. This is the largest contribution to the JES correction.

The response is found to increase with energy, mainly because of the higher response of the calorimeter to photons and electrons relative to hadrons, together with the fact that the fraction of neutral pions in jets increases with jet energy.

The main uncertainties on the corrections are from the energy of the γ and the estimation of dijet contamination of the γ+jet sample.

The DØ γ+jet data have enough events to provide a measurement for the response up to about $E' = 300$ GeV. For a cone $\Delta R = 0.7$ jets, which is used for two and three-body analyses, where the few objects present have high energy, this response is extrapolated to higher energies using simulation.

The second step in the response correction is the measurement of the relative jet response. This uses dijet events with at least one jet in the very central region of the calorimeter ($|\eta_{\text{d}}| \leq 0.4$), while the probe jet can be anywhere in the calorimeter up to $|\eta_{\text{d}}| = 3.6$. This response is called relative as it relates the response in any part of the calorimeter to the central one; the product of the relative response by the absolute response yields the jet response $\hat{R}_{\text{jet}}$.

The response is strongly reduced in the inter-cryostat region, where the scintillator detectors can't provide as good a response as the LAr calorimetry.
The response for γ+jet and dijet events is found to be very different due to the different jet composition. At low energy, dijet events have a larger component of gluons relative to γ+jet events, while at high energies this is reversed.
The main uncertainty to this part of the response comes from the two-dimensional parametrization of the response in terms of E' and η_{d}.

3.4. Out-of-cone energy
The last of the three main corrections takes into account the geometric size of the jet, correcting for (i) the energy of particles from the jet that fall outside the jet cone, (ii) vice versa for the energy of other particles from the main $p\bar{p}$ interaction that accidentally falls inside the jet cone, and (iii) the energy of particles from the jet with very low energy that don't reach the calorimeter due to the magnetic field. The correction does not account for energy lost because of gluon radiation (at parton level) at large angles.

The correction for a jet of cone ΔR (either 0.5 or 0.7), for a given energy and pseudorapidity is:

$$\hat{S}_{\text{jet}}\left(E', \eta\right) = \frac{E^{\text{MC}}_{\text{from-jet}}\left(\delta R < \Delta R\right)}{E^{\text{MC}}_{\text{from-jet}}} + \frac{\alpha}{\beta} \frac{E^{\text{MC}}_{\text{non-jet}}\left(\delta R < \Delta R\right)}{E^{\text{MC}}_{\text{from-jet}}} \tag{5}$$

The first and second ratios describe the fraction of measured energy of particles, respectively from the jet and outside the jet, that fall inside the cone ΔR. As the superscripts suggest, the estimate of the energy in and out of the cone is extracted from Monte Carlo (MC) simulation, using γ+jet events. In fact, this is the only of the main corrections that relies on MC.
The global parameters α and β are simulation-to-data energy scale factors for the two categories of particles (β factor cancels out in the first ratio), that are needed when applying these energy ratios on data. To evaluate the factors, energy profiles of data and MC are compared, and the factors are defined by the energy of particles within an annulus of given radius R from the jet axis: $E\left(R\right) = \sum_{R \leq \delta R < R+0.1} E_{\text{particles}}$. For each E' and η region, a profile is extracted from measured γ+jet data events is extracted. Three profiles are also extracted from MC for the energy from the jet, the energy from outside the jet and the offset energy. The coefficients are then computed from best fit to the relation $E^{\text{data}}_{\text{jet}} = \alpha E^{\text{MC}}_{\text{from-jet}} + \beta E^{\text{MC}}_{\text{non-jet}} + E^{\text{MC}}_{\text{offset}}$, where the offset energy scale can be assumed consistent with data, up to $R \leq 2.0$.

The most relevant uncertainties for this correction are from the simulation of the gluon radiation, the statistical uncertainty on the energy profiles used to estimate α and β, and the purity of the γ+jet sample in data.

4. Uncertainties
Beside the three main components of the JES correction, other minor corrections are also performed, with each of these relying on a variety of assumptions, with many sources of uncertainty identified. Most of these are very small. The largest uncertainty is from the largest correction, namely the energy response (Fig. 6). While most of these uncertainties can be reduced with additional data, the conditions and performances of the detector and the accelerator also change, so that it is more effective to split the correction into different running periods, which reduces the data available for each period.

5. Validation of the corrections
Many analyses strongly rely on the consistency of jet energy scale between data and simulation: in fact, we often measure quantities calibrated with simulation.

We verify the consistency of JES applied on data and simulation (Fig. 7) analysing γ+jet events, integrated in the simulation with additional multijet events that reflect the contamination in the data.

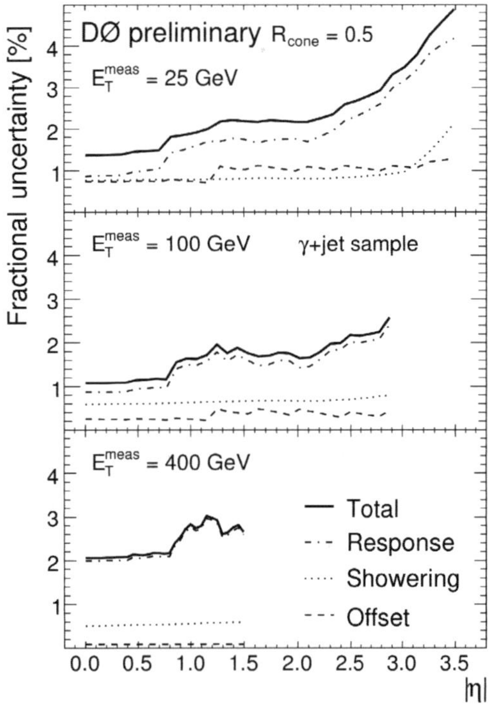

Figure 6. Relative uncertainty on Jet Energy Scale correction, split into the three different main sources, for three measured jet energies (E_T^{meas}). These uncertainties are extracted from ≈ 1.3 fb^{-1} DØ γ+jet data.

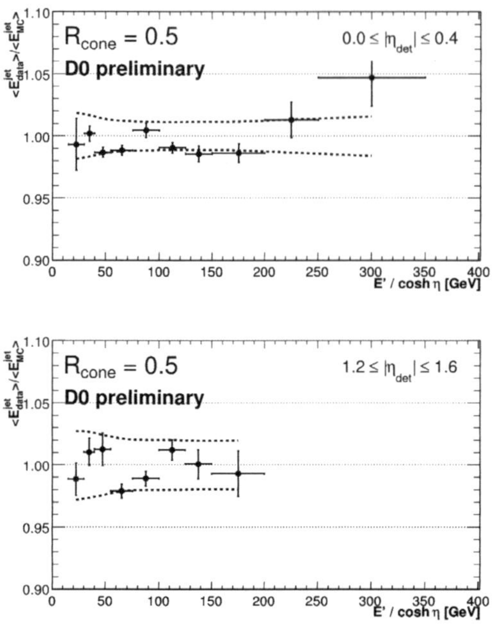

Figure 7. Validation of corrections: ratio between the corrected energy of jets in data and simulation, in two detector regions, as function of the jet energy E'. The difference is statistically consistent within the uncertainty band shown shaded.

6. Summary

At high energy hadron colliders, the Jet Energy Scale is of paramount importance for most analyses, since jets are very common objects in signals and omnipresent in backgrounds. A comprehensive derivation of the Jet Energy Scale corresponds to a separate major analysis, made very complex by the demand of precision of the other analyses that rely on it. There are many corrections that have been ignored in this brief report. The Jet Energy Scale is measured with a precision of the order of 2% in most regions of the calorimeter, with the best precision achieved in the central part of the detector, especially for low-to-medium jet energies, which are the most employed in most physics analyses. The main corrections are extracted directly from data, limiting the propagation of uncertainties from simulation to the measurement.

For this quality to be maintained, corrections must be updated periodically to reflect the changing operating conditions, and to include the newly collected data.

References

[1] V. M. Abazov *et al.* [D0 Collaboration], "The Upgraded D0 detector," Nucl. Instrum. Meth. A **565** (2006) 463 [arXiv:physics/0507191].

[2] S. Abachi *et al.* [D0 Collaboration], "The D0 Detector," Nucl. Instrum. Meth. **A338** (1994) 185-253.

[3] G. C. Blazey *et al.*, "Run II jet physics," arXiv:hep-ex/0005012.

Inclusive jet cross section measurement in ATLAS

Francesc Vives on behalf of the ATLAS Collaboration
IFAE, Universitat Autònoma de Barcelona, Edifici Cn, Campus UAB, E-08193 Bellaterra, Spain

E-mail: Francesc.Vives.Vaque@cern.ch

Abstract. The inclusive jet cross section has been measured in proton-proton collisions at a center-of-mass energy of 7 TeV using the ATLAS detector at the LHC. The full 2010 data set has been used, consisting of a total integrated luminosity of 37 pb^{-1} . The anti-k_t algorithm is used to identify jets with two jet radius parameters, R = 0.4 and R = 0.6. Jet shapes have been measured in order to validate the models for parton shower, fragmentation into hadrons, and the underlying event contributions contained in the event generators. The inclusive jet cross section measurement is presented as a function of jet transverse momentum and rapidity, for jets with transverse momentum from 20 GeV to 1.5 TeV in a rapidity range $|y|$ up to 4.4. The data are compared to expectations based on next-to-leading order QCD corrected for non-perturbative effects, and to predictions obtained using POWHEG. In addition to a validation of the theory in a new kinematic regime, the data also provide sensitivity to parton distribution functions in a region where they are currently poorly constrained.

1. Introduction

The jet cross section measurement is a fundamental test of QCD and any deviation could point to the presence of new physics. It also provides information on the parton distribution functions (PDFs) and the strong coupling.

The first measurement of the inclusive jet cross section in ATLAS was done using an integrated luminosity of about 17 nb^{-1} for jets with $p_T > 60$ GeV and $|y| < 2.8$ [1]. In this article, a measurement of the inclusive jet cross section for jets with $p_T > 20$ GeV and $|y| < 4.4$ [2] is presented, based on 37.3 ± 1.2 pb^{-1} of integrated luminosity collected by ATLAS. The ATLAS detector is described in these proceedings [3].

2. Jet reconstruction and calibration

Jets are reconstructed using the anti-k_t algorithm, with jet radius parameter $R = 0.4$ and $R = 0.6$, run over calorimeter topoclusters [3]. The analysis is based in the full 2010 data set of pp collisions at $\sqrt{s} = 7$ TeV collected by the ATLAS detector, corresponding to an integrated luminosity of about 37 pb^{-1}. Events are considered when the detector components relevant for this analysis were operating at the nominal conditions.

Events are then required to have at least one primary vertex with five or more tracks pointing to it in order to remove backgrounds such as cosmic ray muons. For events containing a jet with $p_T > 20$ GeV, this requirement has an efficiency well above 99%.

Jets with calibrated $p_T > 20$ GeV and $|y| < 4.4$ are selected, and required to pass quality criteria established to reject jets not coming from a proton-proton collision. Fake jets may originate from rare noise bursts in the calorimeter, cosmic rays or beam background. A detailed description of the jet energy calibration can be found in another article of these proceedings [4].

Due to the increase of the instantaneous luminosity during 2010, it became necessary to apply prescales to some triggers used in this analysis in order to limit their trigger rate. For each jet p_T and rapidity bin, a trigger is used to select jets such that is fully efficient ($> 99\%$) and that the prescale is as low as possible. Prescales are taken into account in the luminosity calculation.

3. Unfolding to the particle level

Measurements are corrected for detector effects back to the particle level with a bin-by-bin unfolding procedure, using PYTHIA 6.423 with the AMBT1 tune simulated samples. In order to improve the agreement between the MC and the data, the MC is reweighted in order to scale its prediction, that uses a modified LO PDF, to that of a NLO PDF. The hadronization corrections are derived in each p_T and $|y|$ bin by computing the ratio between the inclusive jet p_T distributions using truth jets (including muons and neutrinos) and using calorimeter simulated jets.

In order to justify this procedure, it is important to validate the QCD description contained in the event generators, such as parton shower cascades, and the fragmentation and underlying event models. This is done in the measurement of jet shapes in proton-proton collisions at $\sqrt{s} = 7$ TeV presented in [5]. This measurement is based on data collected by the ATLAS experiment corresponding to 3 pb^{-1} of total integrated luminosity, and uses jets reconstructed with the anti-k_t algorithm with $R = 0.6$, in the kinematic range 30 GeV$< p_T <$600 GeV and $|y| <2.8$. The events are required to have one and only one reconstructed primary vertex with a z position within 10 cm of the origin of the coordinate system, which suppresses pile-up contributions from multiple proton-proton interactions in the same bunch crossing, beam-related backgrounds and cosmic rays. The differential jet shape in four representative jet p_T bins is shown in Figure 1. The differential jet shape $\rho(r)$ as a function of the distance $r = \sqrt{\Delta y^2 + \Delta\phi^2}$ to the jet axis is defined as the average fraction of the jet p_T that lies inside an annulus of inner radius $r - \Delta r/2$ and outer radius $r + \Delta r/2$ around the jet axis. The dominant peak at small r indicates that the majority of the jet momentum is concentrated close to the jet axis. As expected, jets get narrower as the p_T increases. The data are compared to predictions from HERWIG++, ALPGEN, PYTHIA-Perugia2010, and PYTHIA-MC09. The jet shapes predicted by PYTHIA-Perugia2010 provide a reasonable description of the data, while HERWIG++ predicts broader jets than the data at low and very high p_T. ALPGEN is similar to PYTHIA-Perugia2010 at low p_T, but produces jets significantly narrower than the data at high p_T. PYTHIA-MC09 tends to produce narrower jets than the data in the whole kinematic range under study. Apart from the results reported in [5], jet shapes at calorimeter level have been computed for $|y| >2.8$ in order to validate the MC samples in the forward region.

4. Systematic Uncertainties

- The **JES uncertainty** is the largest contribution to the uncertainty of the inclusive jet cross section measurement. At low p_T it goes from +30%/-20% in the central region to +80%/-50% in the most forward region, and decreases with increasing jet p_T.
- The **unfolding factors** have been recomputed varying the jet p_T in the MC in order to take into account the jet energy and angular resolution, and the differences on the cross section shape in data and in MC. These unfolding factors have been compared to estimate the systematic uncertainty, that is typically between 2% and 5%, except for the lowest p_T bin where it goes up to 20%.
- The uncertainty on the **luminosity** measurement is 3%.

- The efficiency of matching track jets to calorimeter jets has been evaluated in both data and Monte Carlo to estimate the modeling of the calorimeter **jet reconstruction efficiency** in the MC. The difference is smaller than 2% (3%) for jets with $p_T > 20$ GeV (30 GeV), and is taken as a systematic uncertainty.
- In this analysis, jets are selected with **triggers** that are 99% efficient. A conservative 1% uncertainty overall has been considered to account for the trigger inefficiency.
- The systematic uncertainty on the efficiency of the **jet quality criteria** is taken as a systematic uncertainty on the cross section, and it is always below 1%.

5. Results and conclusions

The inclusive jet cross section unfolded to the hadron level as a function of the jet p_T is shown in Figure 2 for different $|y|$ regions up to $|y| < 4.4$. It is measured from 20 GeV to 1.5 TeV and spans up to ten orders of magnitude. The data are compared to NLO predictions corrected for non-perturbative effects. The NLO predictions are obtained from NLOJET++4.1.2 program with CTEQ 6.6 PDFs, and the corrections for non-perturbative effects are derived using samples produced with PYTHIA 6.423 with the AMBT1 tune and applied to NLO predictions. These corrections are obtained bin-by-bin, comparing the cross section with and without hadronization and underlying event. The correction is dominated by the underlying event at low p_T (1.5 at 20 GeV), and tends to 1 as the p_T increases. The uncertainty on the NLO predictions is obtained adding in quadrature uncertainties from the PDFs, the choice of factorization and renormalization scales, and the value of the strong coupling constant. It is combined with the uncertainty on the non-perturbative corrections, estimated from the maximum difference of the corrections with AMBT1 with respect to other PYTHIA tunes and to HERWIG++. The total uncertainty in the predictions is typically between 10% and 20%.

Figure 3 shows the ratio of the measured cross section in data and the theory prediction, that are in agreement within uncertainties. The cross section in data is lower than the cross section predicted by NLO predictions in the forward region and at high p_T.

In Figure 4, data are compared to results using CTEQ 6.6, MSTW 2008, NNPDF 2.1, and HERAPDF 1.5. All cross sections are normalized to that obtained with CTEQ 6.6. Predictions using MSTW 2008, NNPDF 2.1, and HERAPDF 1.5 are closer to data than those using CTEQ 6.6, but all have the tendency to produce higher cross sections than data at high p_T in the forward region. This may be due to the fact that PDFs are currently poorly constrained in this kinematic region.

Finally, Figure 5 shows the comparison of data and POWHEG predictions with MSTW 2008. POWHEG is showered with both PYTHIA and HERWIG. Data are also compared to NLO predictions with MSTW 2008, used to normalize both data and POWHEG predictions. Even if POWHEG predictions agree with data within uncertainties, they tend to produce larger cross section at low p_T and, mainly POWHEG+HERWIG, smaller cross sections at high p_T. POWHEG predictions are different depending on whether they are showered with PYTHIA or HERWIG.

References

[1] The ATLAS Collaboration, Measurement of inclusive jet and dijet cross sections in proton-proton collisions at 7 TeV center-of-mass energy, with the ATLAS detector, Eur. Phys. J. C **71** 2 1512 (2011)

[2] The ATLAS Collaboration, Measurement of inclusive jet and dijet cross sections in proton-proton collision data at 7 TeV center-of-mass energy using the ATLAS detector, ATLAS-CONF-2011-047 (2011): http://cdsweb.cern.ch/record/1338578

[3] Peter Loch, Jet measurements in ATLAS, 2011 (these proceedings)

[4] Doug Schouten, In-situ measurements of the jet energy scale in ATLAS, 2011 (these proceedings)

[5] The ATLAS Collaboration, Study of jet shapes in inclusive jet production in pp collisions at $\sqrt{s} = 7$ TeV using the ATLAS detector, Phys. Rev. D **83** 052003 (2011)

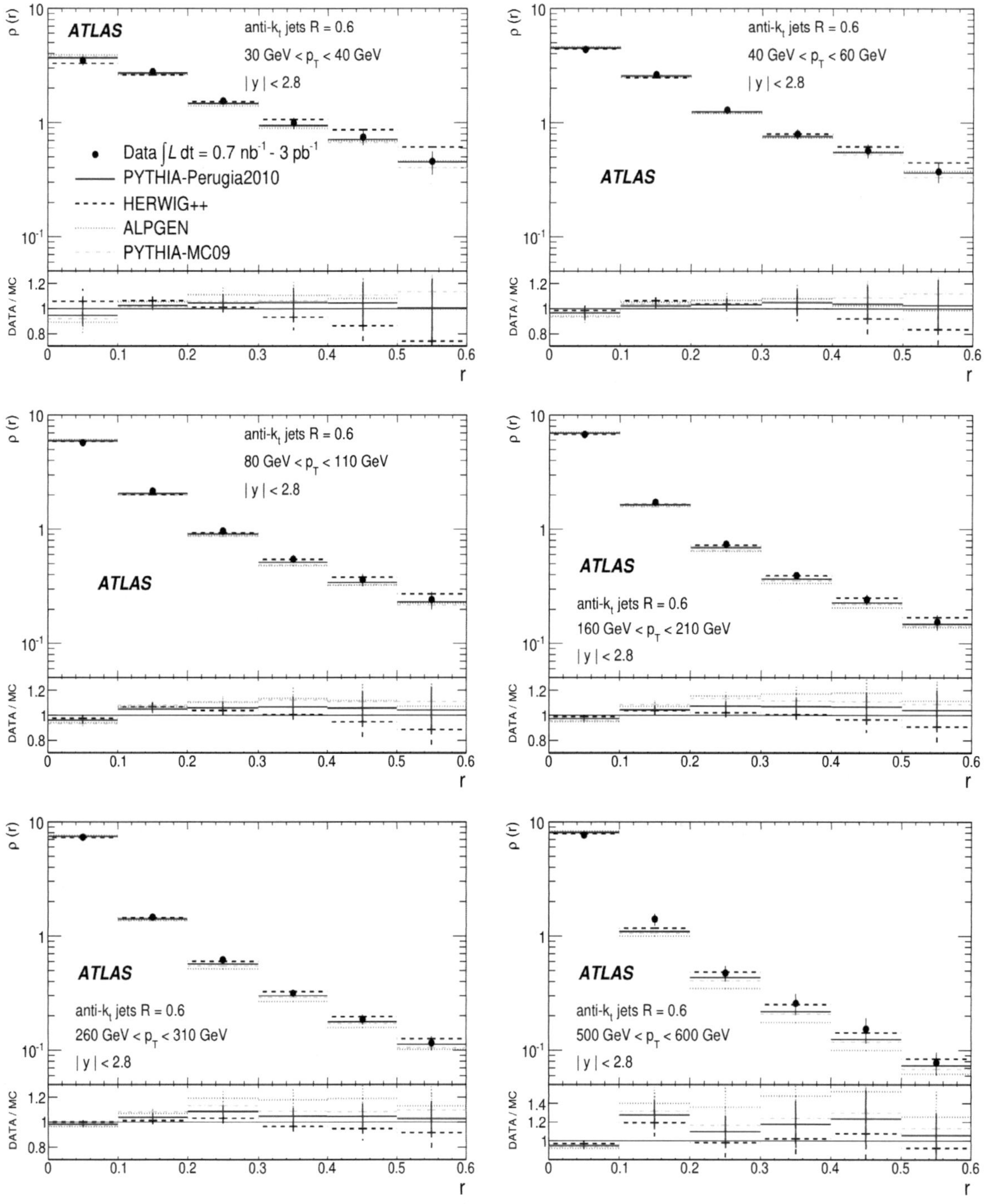

Figure 1. The measured differential jet shape, $\rho(r)$, in inclusive jet production for jets with $|y| < 2.8$ in six representative p_T bins. Error bars indicate the statistical and systematic uncertainties added in quadrature. The predictions of PYTHIA-Perugia2010 (solid lines), HERWIG++ (dashed lines), ALPGEN interfaced with HERWIG and JIMMY (dotted lines), and PYTHIA-MC09 (dashed-dotted lines) are shown for comparison.

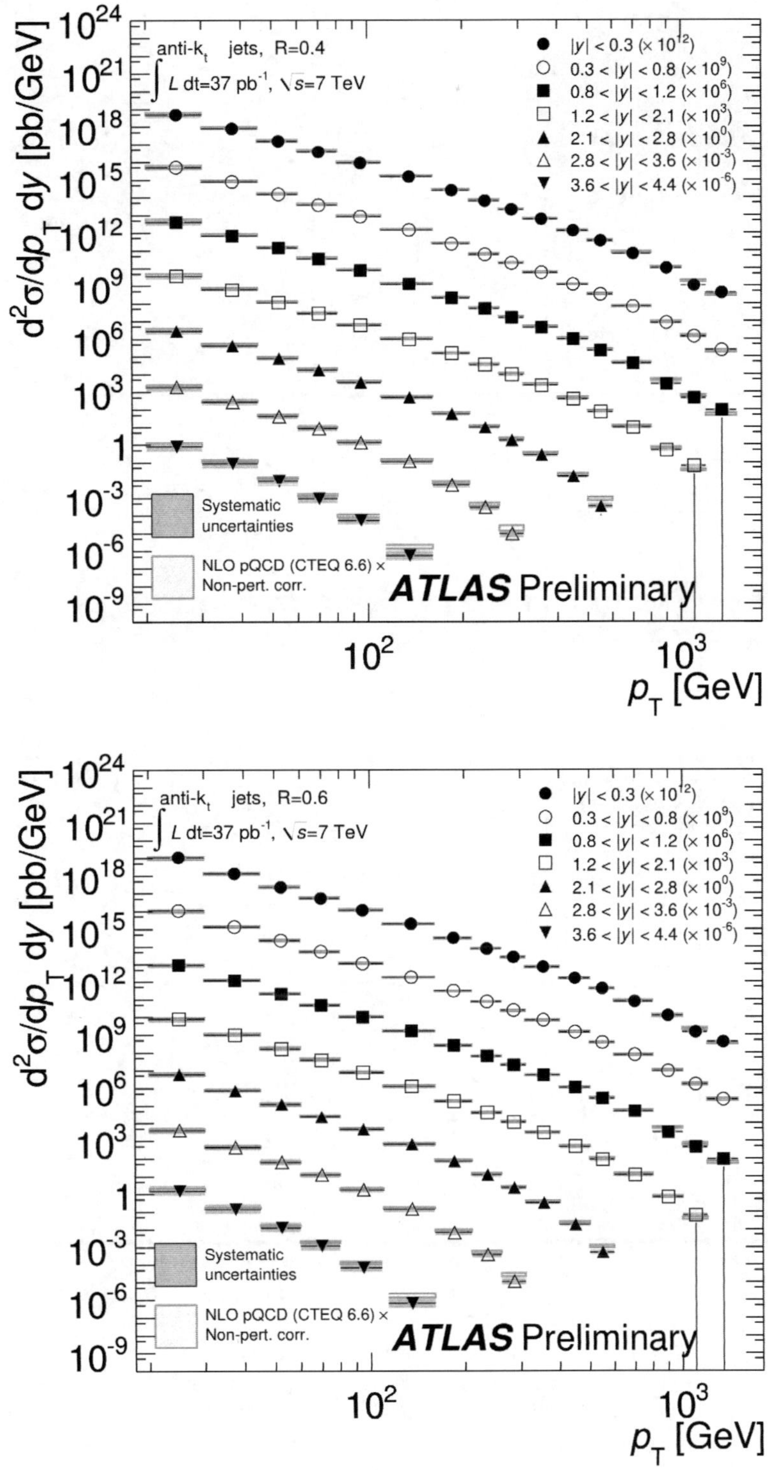

Figure 2. Inclusive jet cross section as a function of p_T in different regions of $|y|$ for jets identied using the anti-k_t algorithm with R = 0.4 (top) and R = 0.6 (bottom). For displaying purposes, the cross sections are multiplied by the factors indicated in the legend. NLO QCD calculations with additional non-perturbative corrections are overlaid on top of the data. The error bars indicate the statistical uncertainty on the measurement, and the shaded band indicates the quadratic sum of the systematic uncertainties. The data points do not include an overall uncertainty of 3% due to the luminosity measurement.

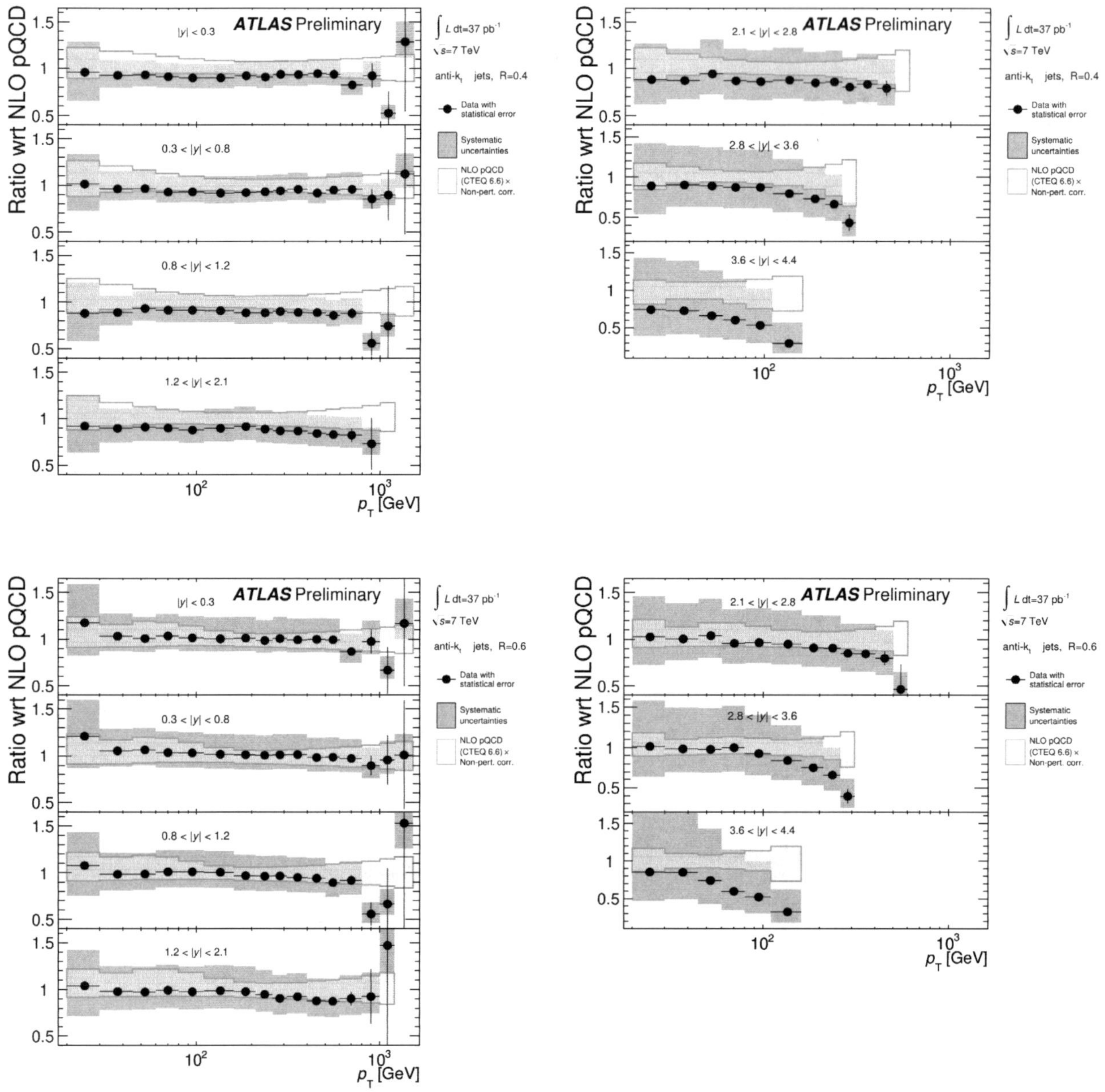

Figure 3. Inclusive jet cross section as a function of p_T in different regions of $|y|$ for jets identied using the anti-k_t algorithm with R = 0.4 (top) and R = 0.6 (bottom). The plots show the ratio of the data to the theoretical prediction. The error bars indicate the statistical uncertainty on the measurement, and the shaded band indicates the quadratic sum of the systematic uncertainties. The data points do not include an overall uncertainty of 3% due to the luminosity measurement.

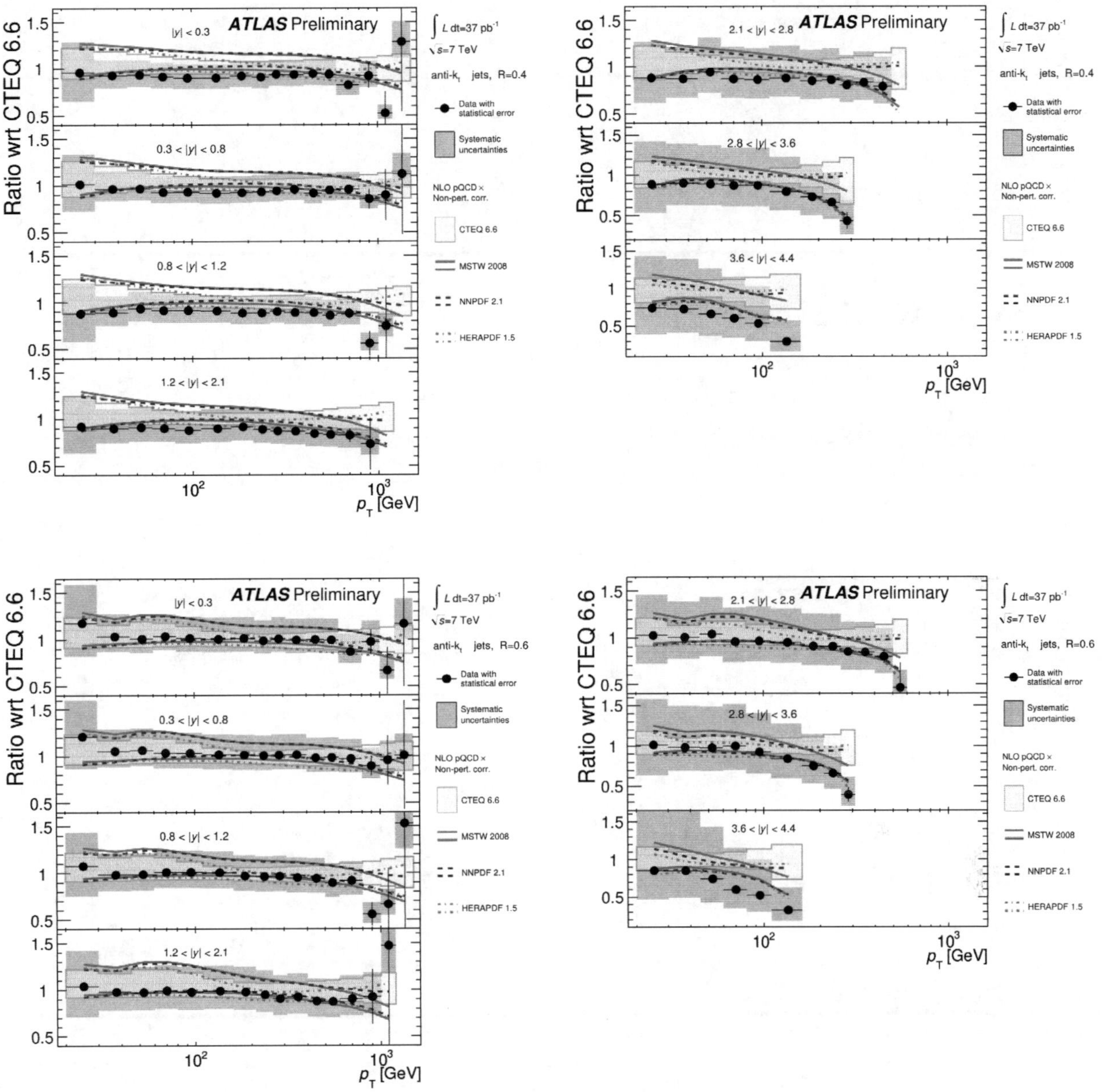

Figure 4. Inclusive jet cross section as a function of p_T in different regions of $|y|$ for jets identied using the anti-k_t algorithm with R = 0.4 (top) and R = 0.6 (bottom). The theoretical error bands obtained by using different PDF sets (CTEQ 6.6, MSTW 2008, NNPDF 2.1, HERA 1.5) are shown. The data points and the error bands are normalized to the theoretical estimates obtained by using the CTEQ 6.6 PDF set.

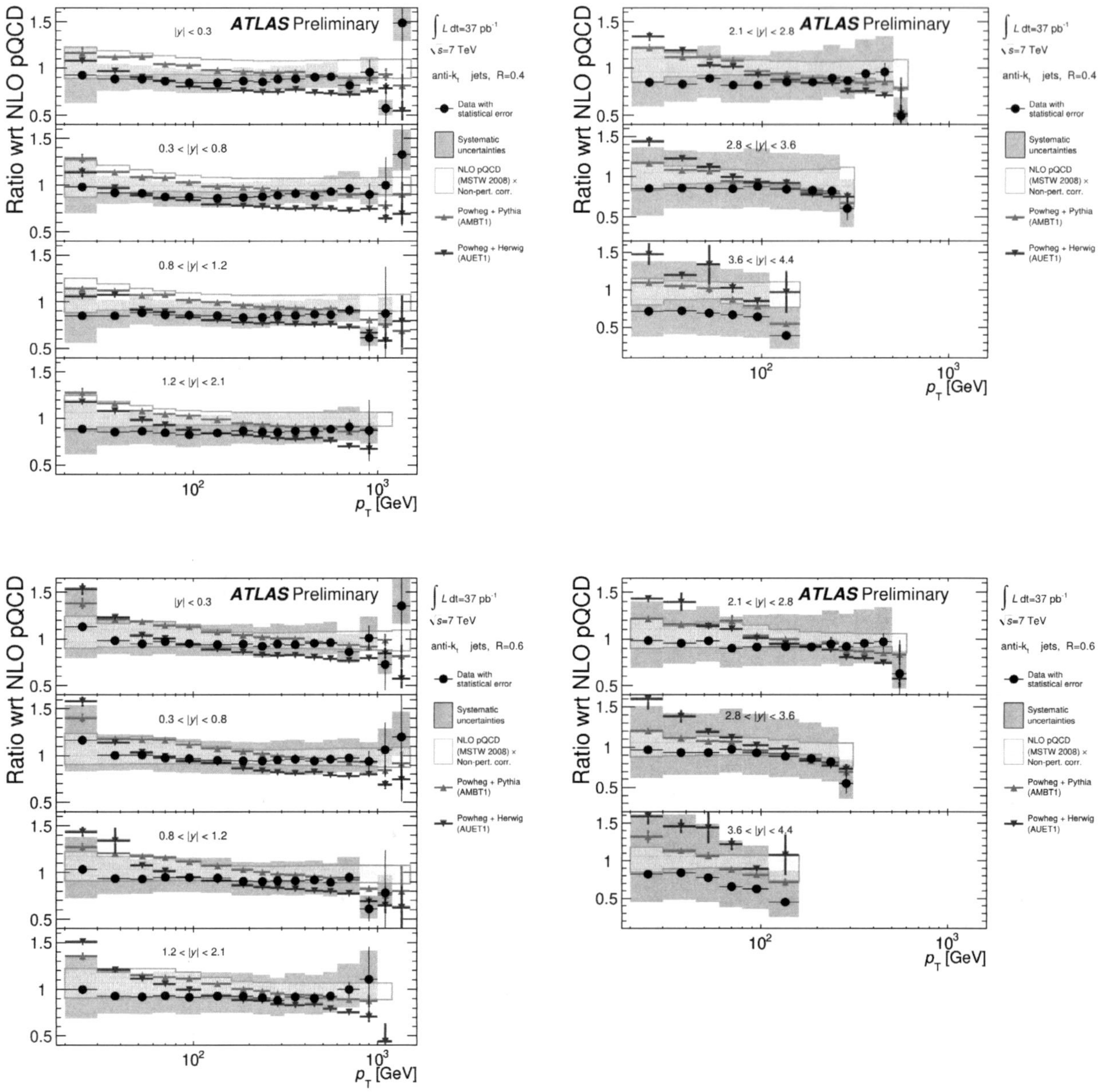

Figure 5. Inclusive jet cross section as a function of p_T in different regions of $|y|$ for jets identied using the anti-k_t algorithm with R = 0.4 (top) and R = 0.6 (bottom). The ratios of the POWHEG predictions showered by PYTHIA and HERWIG and the data to the NLO predictions corrected for the non-perturbative effects is shown. The ratio shows only the statistical uncertainty on the POWHEG prediction. The total systematic uncertainties on the theory and measurement are shown.

Measurement of multi-jet cross-sections in proton-proton collisions at 7 TeV center-of-mass energy in ATLAS

Zinonas Zinonos - on behalf of the ATLAS Collaboration

Università di Pisa & INFN Sezione di Pisa, Italia

E-mail: zenon@cern.ch

Abstract. Inclusive multi-jet production is studied using the ATLAS detector for proton-proton collisions with a center-of-mass energy of 7 TeV at the Large Hadron Collider at CERN. The data sample corresponds to an integrated luminosity of 2.4 pb^{-1}, using the first proton-proton data collected by the ATLAS detector in 2010. Results on multi-jet cross sections are presented and compared to both leading-order plus parton-shower Monte Carlo predictions and next-to-leading-order QCD calculations.

1. Introduction

Multi-jet cross sections serve as one of the main observables in high-energy particle physics. Events containing multiple jets in the final state are prolific and provide a good testing ground for the theory of the strong interaction, quantum chromodynamics (QCD). At high transverse momentum (high p_T), the production of jets is modeled by QCD as the hard scattering of partons and the subsequent parton showering, followed by a hadronization process. Within the framework of QCD, the jet energy is related to the energy of partons produced in the hadron collisions. Consequentially, the study of energy distributions for multi-jet events provides a fundamental and direct test of QCD at hadron colliders.

This is the first study [2] performed of multi-jet events from proton-proton collisions at 7 TeV center-of-mass energy, using the ATLAS detector at the Large Hadron Collider (LHC) at CERN. The data sample used for the analysis was collected between April 10 and August 30 of 2010 and represents a total integrated luminosity of 2.4 pb^{-1} [1]. Approximately half a million events with at least two jets in the final state are selected using this data sample.

Two primary motivations for the multi-jet study are to evaluate how robust leading-order (LO) QCD calculations are in representing the high jet multiplicity events, relevant as backgrounds in high-energy searches, and to test next-to-leading-order perturbative QCD (NLO pQCD) calculations. For the leading-order comparisons, events with up to six jets in the final state are studied, and for the next-to-leading-order perturbative QCD study the focus is on three-jet events and their comparison to two-jet events.

2. The ATLAS Detector

ATLAS is a general-purpose detector surrounding Interaction Point 1 of the LHC and its architecture is based on a system of solenoidal and toroidal magnets [3].

High-energy particles produced in collisions initially pass through an inner tracking system embedded in a strong solenoidal magnetic field. The inner tracker covers a pseudorapidity $|\eta| < 2.5$ and has full coverage in azimuth. In the barrel region, it is made of three main components arranged in concentric layers, all of which are immersed in a 2 Tesla field provided by the inner solenoid magnet. Three layers of silicon pixel detectors provide a two-dimensional hit position very close to the interaction point. Silicon microstrip detectors are then used in the next four layers, providing excellent position resolution for charged particles. A transition-radiation detector is the outermost component of the tracker, with coarser position resolution with respect to the silicon, but giving a large lever-arm for track reconstruction in addition to particle identification capabilities. The ATLAS calorimeter is also composed of many sub-detectors. The electromagnetic calorimeter ($|\eta| < 3.2$) is a high-granularity sampling detector in which the sensitive medium is liquid argon (LAr). The hadronic calorimeters are divided into three sections: a tile scintillator/steel calorimeter in used in both the barrel ($|\eta| < 1.0$) and extended barrel cylinders ($0.8 < |\eta| < 1.7$); the hadronic endcap covers the region $1.5 < |\eta| < 3.2$ and consists of LAr/copper calorimeter modules. Finally the forward calorimeter measures both electromagnetic and hadronic energy in the range $3.2 < |\eta| < 4.9$ using LAr/copper and LAr/tungsten modules. The total coverage of the ATLAS calorimeters is therefore $|\eta| < 4.9$. A muon spectrometer, which is designed to identify muons and measure both their trajectories and momenta with high accuracy: the design momentum resolution is 10% at momenta transverse to the beam line (p_T) of 1 TeV. The muon spectrometer, comprises three toroidal magnet systems consisting of eight coils each with a bending power $\int Bdl = 1 - 7.5$ Tm.

The ATLAS trigger system employs three trigger levels, of which only the hardware-based first level trigger is used in this analysis. Events are selected using the calorimeter-based jet trigger. The first level jet trigger [4] uses coarse detector information to identify areas in the calorimeter where energy deposits above a certain threshold occur. A simplified jet finding algorithm based on a sliding window of size $\Delta\phi \times \Delta\eta = 0.8 \times 0.8$ is used to identify these areas. This algorithm uses coarse calorimeter towers with a granularity of $\Delta\phi \times \Delta\eta = 0.2 \times 0.2$ as inputs.

3. Cross Section Definitions and Kinematics

The anti-k_t algorithm [5, 6] with full four-momentum recombination is used to identify jets.

For high multiplicity studies, namely up to six jets in an event, the resolution parameter in the jet reconstruction is fixed to $R = 0.4$ to contend with the limited phase space and to reduce the impact of the underlying event [7] in the jet energy determination. For testing NLO pQCD calculations, where the study focuses on three-jet events, a resolution parameter of $R = 0.6$ is preferred, since a larger value of R is less sensitive to theoretical scale uncertainties. The anti-k_t algorithm was chosen because it can be implemented in the NLO pQCD calculation, is infra-red safe to all orders and produces jets with a simple geometrical shape.

Jet measurements are corrected for all experimental effects and refer to the particle-level final state. At the particle level, jets are built using all final-state particles with a proper lifetime longer than 10 ps, including muons and neutrinos from hadronic decays.

Cross sections are calculated in bins of inclusive jet multiplicity, meaning that an event is recorded in a jet multiplicity bin if it contains a number of jets that is equal to or greater than that multiplicity. For example, an event with three reconstructed jets will be counted both in the two-jet and three-jet multiplicity bin. Inclusive multiplicity bins are used because they are stable in the pQCD fixed-order calculation, unlike exclusive bins. Only jets with $p_T \geq 60$ GeV and $|y| \leq 2.8$ are counted in the measurement. These cuts are chosen to ensure that the jets are reconstructed with high efficiency. The leading jet is further required to have $p_T \geq 80$ GeV to stabilize the NLO pQCD calculations [8].

4. Theoretical Predictions

Measurements are compared to pQCD calculations at leading order and next-to-leading order. The different predictions are all normalized to the measured inclusive two-jet cross section and then used for shape comparisons.

For the leading-order analysis, ALPGEN [9] is used to generate events with up to six partons in the final state using the leading-order set of proton PDFs CTEQ6L1 [10]. ALPGEN is interfaced to HERWIG/JIMMY [12, 13] or PYTHIA [11] Monte Carlo generators in order to sum leading logarithms to all orders in the parton-shower approximation and to include non-perturbative effects such as hadronization and the underlying event. This was done by evaluating the ratio of the cross-section with and without hadronisation and underlying event simulation using different leading-logarithmic parton-shower Monte Carlo programs and then multiplying the NLO theory distributions by this correcting factor.

SHERPA [14] with its default parameters and renormalization scale scheme from version 1.2.3 is also used to generate events with up to six partons in the final state. This provides an independent matrix-element calculation with a different matching scheme between the matrix element and the parton shower.

The PYTHIA 6.421 [15] event generator is also compared to the data to study the limitations of leading-order $2 \to 2$ matrix-element calculations. This generator implements a leading-order matrix-element calculation for $2 \to 2$ processes, p_T-ordered parton showers, an underlying-event model for multiple-parton interactions and the Lund string model for hadronization.

For the purpose of understanding detector effects, the particles generated in the leading-order Monte Carlo generators are passed through a full simulation of the ATLAS detector and trigger [16] based on GEANT4 [17]. Additional proton-proton collisions are added to the hard scatter in the simulation process to reproduce realistic LHC running conditions. Events and jets are selected using the same criteria in data and Monte Carlo simulations.

For the next-to-leading-order pQCD study, the calculation implemented in NLOJet++ [18] is used. The renormalization and factorization scales are varied independently by a factor of two in order to estimate the impact of higher order terms not included in the calculation. An additional requirement that the ratio of the renormalization and factorization scales did not differ by more than a factor of two was imposed.

The NLOJet++ program implements only a matrix-element calculation, lacking a parton-shower interface. Therefore, it does not account for non-perturbative effects such as the hadronization and underlying event. To compare to particle-level measurements, a correction factor is applied which takes the next-to-leading-order pQCD calculations to the particle level.

5. Jet Reconstruction and Calibration

Jets are reconstructed at the electromagnetic scale using the anti-k_t algorithm. The input objects to the jet algorithm are three-dimensional topological clusters [19] built from calorimeter cells. The four-momentum of the uncalibrated, EM-scale jet is defined as the sum of the four-momenta of its constituent calorimeter energy clusters. Additional energy due to multiple proton-proton interactions within the same bunch crossing (pile-up) is subtracted by applying a correction derived using minimum bias data. The energy and the position of the jet are corrected for instrumental effects such as dead material and non-compensation. This jet energy scale (JES) correction is calculated using isolated jets in the Monte Carlo simulation as a function of the transverse momentum and pseudorapidity of the reconstructed jet [20].

6. Offline Selection

6.1. Trigger

A set of ATLAS first level (level-1) multi-jet triggers is used to select events for the analysis. Multi-jet triggers require several jets reconstructed with a level-1 sliding window algorithm. All multi-jet triggers are symmetric, meaning that each trigger had one particular energy threshold that applied equally to all jets in an event. Only two-jet and three-jet triggers were needed for the analysis.

The trigger efficiencies have been calculated using a bootstrap technique in data as well in Monte Carlo simulated events. Figure 1 shows the efficiency for the third leading jet to fire the three-jet trigger as a function of the reconstructed jet p_T for jets of R = 0.4. The trigger becomes fully efficient when at least one jet with offline $p_T > 60$ GeV is found in the fired event.

The event-level efficiency as a function of the closest distance between two selected $R = 0.4$ offline jets for events selected using the three-jet trigger is also shown in Figure 2, to probe possible topological dependences in the trigger. A dependence at low ΔR is observed, where $\Delta R = \sqrt{\Delta\phi^2 + \Delta y^2}$ represents the minimum separation between selected jets in the event. The dependence on ΔR is well described by the Monte Carlo simulation. Such an inefficiency due to the nearby jet activity is taken into account in the data correction (Section 7). However, this effect is not observed in the analysis of jets reconstructed using the anti-k_t algorithm with resolution parameter R = 0.6.

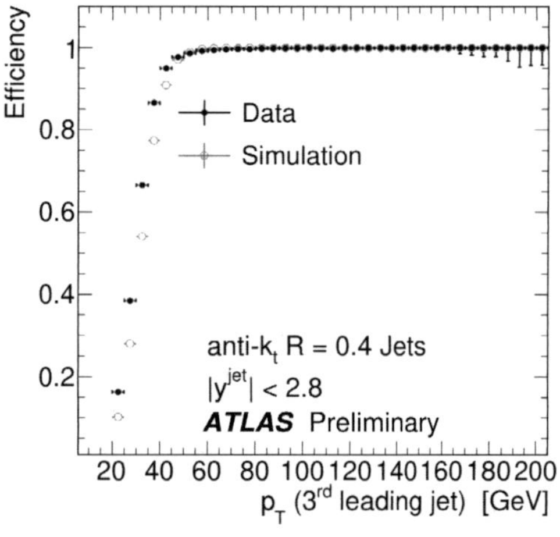

Figure 1. Jet trigger efficiency for the third leading jet as a function of p_T for anti-k_t jets with $R = 0.4$.

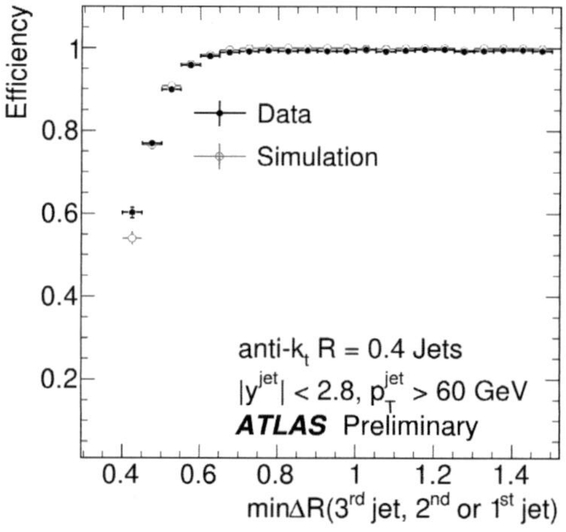

Figure 2. Jet trigger efficiency as a function of the minimum separation ΔR between the two closest jets.

All events falling in the three-jet inclusive multiplicity bin or higher are selected using the three-jet trigger with a jet threshold of 10 GeV on the level-1 jet objects. In order to select events in the two-jet *inclusive* multiplicity bin, several two-jet triggers were used. Three two-jet triggers with symmetric transverse energy thresholds of 10, 15 and 30 GeV were combined independently, weighted by the luminosity associated with each trigger. More specifically, the three triggers with thresholds of 10, 15 and 30 GeV covered the ranges of second leading jet p_T of 60-80 GeV, 80-110 GeV and greater than 110 GeV, respectively.

6.2. Event Selection

To reject events due to cosmic-ray muons and other non-collision backgrounds, events are required to have at least one primary vertex that is consistent with the beam spot position and that has at least five tracks associated to it. The efficiency for collision events to pass these vertex requirements, as measured in a sample of events passing all selections of this analysis, is well over 99%.

6.3. Jet Selection Criteria

Jets considered in the analysis are selected using the following kinematic and data quality selection criteria:

(i) An event must contain at least one jet with $|y| \leq 2.8$ and a p_T greater than or equal to 80 GeV.

(ii) Jets are required to have $|y| \leq 2.8$ and $p_T \geq 60$ GeV in order to be counted.

(iii) Jets are only accepted if at least 70% of the total p_T of the associated charged tracks comes from the event vertex. Overall, this cut reduces the two-jet cross section by 0.4%, and its effect increases with jet multiplicity. The cut reduces the six-jet cross section by 3.4%. All observables show a negligible dependence on the number of reconstructed primary vertices when this cut is applied. Jets with no charged particle content are accepted.

(iv) Only events with a minimum of two selected jets are used in the analysis.

For a total integrated luminosity of 2.4 pb^{-1}, approximately 500,000 multi-jet events survived the selection cuts.

7. Data Correction

A correction is needed to compare the measurements to theoretical predictions. The correction, which accounts for trigger inefficiencies, detector resolutions and other detector effects that affect the jet counting, is performed in a single step using a bin-by-bin unfolding method calculated from Monte Carlo simulations. For each measured distribution, the corresponding Monte Carlo simulation cross section using truth jets, as defined in Section 5, is evaluated in the relevant bins, along with the equivalent distributions obtained after the application of detector simulation and analysis cuts. The ratio of the true to the simulated distributions provides the multiplicative correction factor (unfolding factor) to be applied to the measured distributions.

8. Uncertainty on the Jet Energy Scale

The uncertainty in the jet energy scale is the dominant systematic uncertainty for most of the measurements performed in analysis. The rather steeply falling cross sections as a function of jet p_T implies that even a relatively small uncertainty in the determination of the jet p_T translates into a substantial change in the cross sections as events migrate up or down the steeply falling curve.

The standard jet energy scale (JES) uncertainty in ATLAS [21] has been calculated using jets from a dijet sample without near-by activity in the calorimeter. The JES uncertainty in the central calorimeter region ($\eta < 0.8$) is lower than 2.5% for jets with transverse momentum $60 < p_T < 800$ GeV, and less than 4.6% for the full p_T-range $p_T > 20$ GeV. In the endcap ($0.8 < |\eta| < 2.8$) and forward ($2.8 < |\eta| < 4.5$) regions, the uncertainty for jets with $p_T > 50$ GeV is below 4% and 6% respectively.

For a multi-jet analysis, a set of additional systematic uncertainties are considered. The uncertainties arise from the different calorimeter response to jets of different flavors as well as effects that are sensitive to the available phase space for multi-jet events. These different effects are taken into account in the overall uncertainty of the JES.

9. Results

Figure 3 shows the results for the cross section as a function of the inclusive jet multiplicity. The data are compared to leading-order Monte Carlo simulations (ALPGEN+HERWIG with AUET1 tune, ALPGEN+PYTHIA tuned with MC09′, PYTHIA with AMBT1 tune and SHERPA with its default settings) normalized to the measured inclusive two-jet cross section. The measurement systematics are dominated by the jet energy scale uncertainty and range from 10-20% at low multiplicities to almost 30-40% at high multiplicities. The Monte Carlo simulation predictions fall on the measured results across the full inclusive multiplicity spectrum.

A study that reduces significantly the impact of systematic uncertainties is the ratio of the n-jet to $(n-1)$-jet cross section as a function of multiplicity. Figure 4 shows the results for such a study. Both the unfolding and the jet energy scale uncertainties contribute comparably to the total systematic uncertainty, whereas the statistical uncertainties are smaller than the systematic uncertainties, and negligible in most bins. All Monte Carlo simulations agree well with the data, yet there is a noticeable spread in their predictions. Differences at the level of 15% are observed between PYTHIA AMBT1 and ALPGEN+PYTHIA MC09′ [22] ATLAS tunes in the first bin. These differences most likely arise from the difference between the ME+PS and the pure parton-shower Monte Carlo simulations.

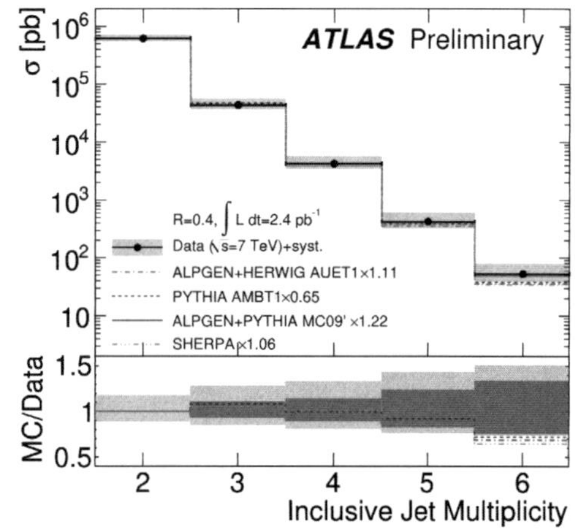

Figure 3. Total inclusive jet cross section as a function of multiplicity. The orange error bands correspond to the systematic uncertainties in the measurement, excluding the luminosity uncertainty.

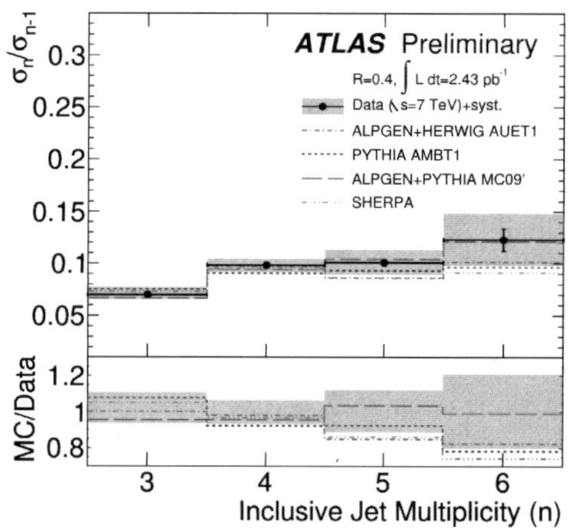

Figure 4. Ratio of the n-jet cross section to the $(n-1)$-jet cross section for values of n varying from three to six. Systematic uncertainties on the cross section ratios are shown as an error band.

The differential cross section for multi-jet events as a function of the jet p_T is useful for characterizing the kinematic features. Figures 5-6 present the p_T-dependent differential cross sections for the leading and second leading jet in multi-jet events. The systematic uncertainty in the measurement is 10-20% across p_T and increasing up to 30% for the fourth leading jet differential cross section. The jet energy scale systematic uncertainty remains the dominant uncertainty in the measurement. All Monte Carlo simulations agree reasonably well with the data. The PYTHIA AMBT1 Monte Carlo simulation predicts a somewhat steeper slope, compared to the data, as a function of the leading jet p_T and of the second leading jet p_T. The effect is most noticeable in the highest p_T range.

NLO pQCD calculations of the three-to-two-jet cross section ratio were performed as a function of the sum of the p_T of the two leading jets, $H_T^{(2)}$. The NLO pQCD calculation for the ratio as a function of this specific observable was found to give the smallest theoretical scale uncertainty and is, therefore, most sensitive to input parameters such as the α_S fit.

Figure 7 shows a comparison of the measurement to NLO pQCD calculations with the MSTW 2008 NLO PDF [23] set for R = 0.6 jets. Scale uncertainties of the NLO pQCD calculations are found to be larger for jets with R = 0.4 than with R = 0.6 [2]. The theoretical uncertainty of the

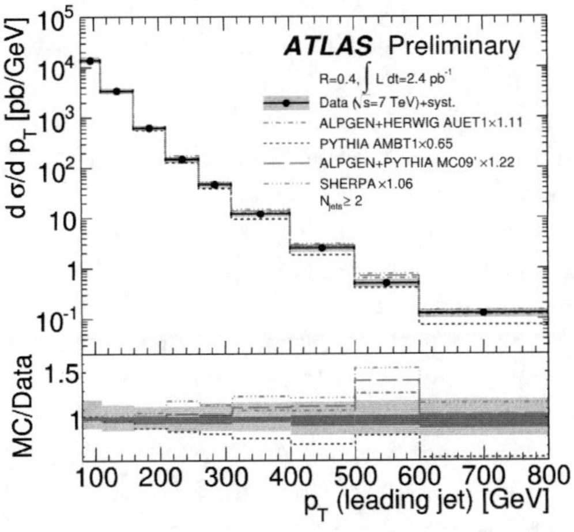

Figure 5. Differential cross section as a function of leading jet p_T for events with $N_{\text{jets}} \geq 2$ jets.

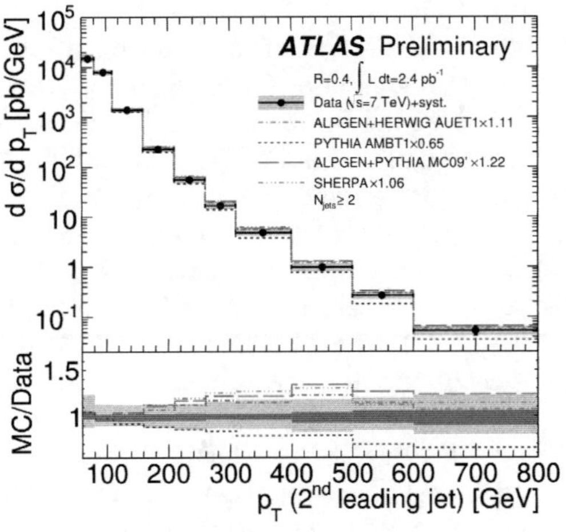

Figure 6. Differential cross section as a function of sub-leading jet p_T for events with $N_{\text{jets}} \geq 2$ jets.

NLO pQCD calculations is comparable to the measurement uncertainties of data. A comparison of the same measurement to leading-order Monte Carlo simulations is given in Figure 8 showing a very good agreement.

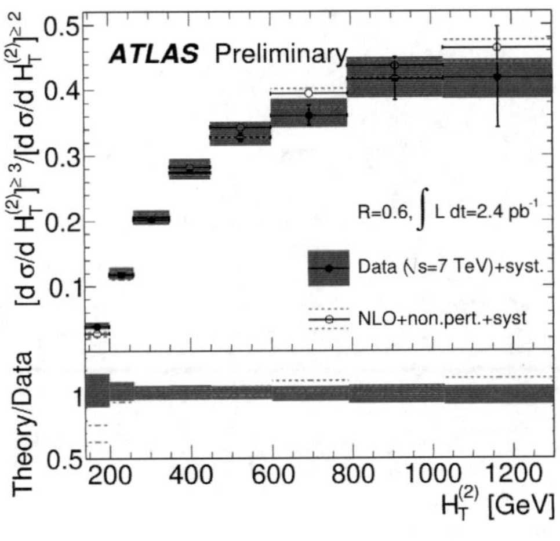

Figure 7. Three-to-two-jet differential cross-section ratio as a function of the sum of the p_T of the two leading jets using $R = 0.6$. The data results are compared to NLO pQCD calculations.

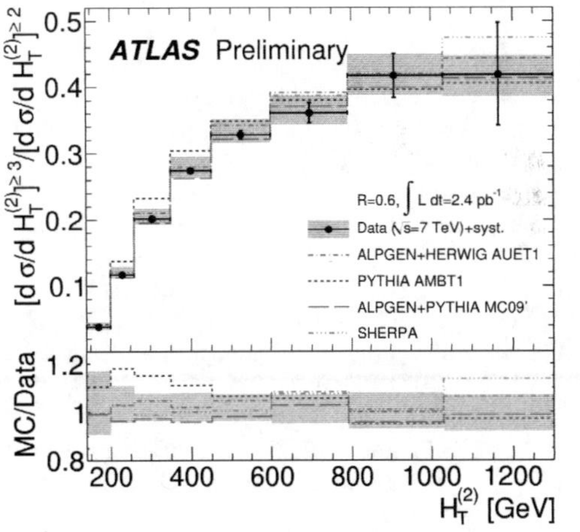

Figure 8. Three-to-two-jet differential cross-section ratio as a function of $H_T^{(2)}$ for jet with a resolution parameter R = 0.6. Data measurements are compared to several leading-order Monte Carlo simulations.

10. Conclusions

A first dedicated study of multi-jet events has been performed using the ATLAS detector at a center-of-mass energy of 7 TeV with an integrated luminosity of 2.4 pb^{-1} [2].

For events containing two or more jets, good agreement is found between data and leading-order Monte Carlo simulations with parton-shower tunes that describe adequately the ATLAS $\sqrt{s} = 7$ TeV underlying-event data.

All models reproduce the main features of the multi-jet data. ALPGEN, which contains higher multiplicity tree-level matrix elements matched to parton showers, generally describes the shapes well, whether the HERWIG or PYTHIA parton showers are used. Similarly, SHERPA describes the data well. However, PYTHIA, which contains a $2 \rightarrow 2$ leading-order matrix element augmented by parton showers, shows only a marginal agreement with the data.

A measurement of the three-to-two-jet cross section ratio as a function of the the sum of the two leading jet p_Ts has also been performed and is described by ALPGEN, SHERPA and a NLO pQCD calculation. Comparisons with NLO pQCD calculations may be useful for constraining parameters, such as parton distribution functions or the value of the strong coupling constant, α_S, given that the systematic uncertainties from the measurement are comparable to the theoretical uncertainties.

References

[1] G. Aad *et al.*, ATLAS Collaboration, "Luminosity Determination in pp Collisions at sqrt(s)=7 TeV Using the ATLAS Detector at the LHC", Eur. Phys. J. **C71**, 1630 (2011), arXiv:1101.2185 [hep-ex].

[2] G. Aad *et al.*, ATLAS Collaboration, "Measurement of multi-jet cross sections in proton-proton collisions at a 7 TeV center-of-mass energy," arXiv:1107.2092 [hep-ex].

[3] ATLAS Collaboration, "The ATLAS Experiment at the CERN Large Hadron Collider", JINST 3 (2008) S08003.

[4] R. Achenbach et al., "The ATLAS Level-1 Calorimeter Trigger", Journal of Instrumentation 3, 2008, P03001.

[5] M. Cacciari, P. G. Salam, and G. Soyez, "The anti-kt jet clustering algorithm", JHEP 04 (2008) 063.

[6] M. Cacciari, G. P. Salam, G. Soyez, http://fastjet.fr/

[7] ATLAS Collaboration, "Measurement of underlying event characteristics using charged particles in pp collisions at $\sqrt{s} = 900 GeV$ and 7 TeV with the ATLAS detector," arXiv:1012.0791 [hep-ex].

[8] S. Frixione and G. Ridolfi, "Jet photoproduction at HERA", Nucl. Phys. B507 (1997).

[9] M. L. Mangano et al., "ALPGEN, a generator for hard multiparton processes in hadronic collisions", JHEP 07 (2003) 001.

[10] J. Pumplin et al., "New generation of parton distributions with uncertainties from global QCD analysis", JHEP 07 (2002) 012.

[11] R. Corke, "Multiple Interactions in Pythia 8", arXiv:0901.2852 [hep-ph].

[12] G. Corcella et al., "HERWIG 6.5 release note", arXiv:hep-ph/0210213.

[13] ATLAS Collaboration, "First tuning of HERWIG/JIMMY to ATLAS data", ATLAS Note ATLAS-PHYS-PUB-2010-014 (2010).

[14] T. Gleisberg et al., "Event generation with SHERPA 1.1", JHEP 02 (2009) 007.

[15] T. Sjostrand et al., "High-energy physics event generation with PYTHIA 6.1", C. Phys. C. 135 (2001) 238.

[16] ATLAS Collaboration, "The ATLAS Simulation Infrastructure", Eur. Phys. J. C70 (2010) 823, arXiv:1005.4568.

[17] S. Agostinelli et al., "Geant simulation", Nucl. Instr. and Meth. A 506 (2003) 250.

[18] Z. Nagy, "Next-to-Leading order calculation of three jet observables in hadron hadron collisions", Physical Review D (2003) 094002.

[19] W. Lampl et al., "Calorimeter Clustering algorithms: Description and Performance", ATLAS-LARG-PUB-2008-002, 2008.

[20] ATLAS Collaboration, "Jet energy scale and its systematic uncertainty in proton-proton collisions at $\sqrt{s} = 7$ TeV with ATLAS 2010 data", ATLAS Note ATLAS-CONF-2011-032 (2011).

[21] ATLAS Collaboration, "Final jet energy scale and its systematic uncertainty for jets produced in proton-proton collisions at $\sqrt{s} = 7$ TeV and measured with the ATLAS detector for the 2010 dataset", ATLAS Note ATLAS-CONF-2011-032 (2011).

[22] ATLAS Collaboration, "ATLAS Monte Carlo Tunes for MC09", ATLAS Note ATLAS-PHYS-PUB-2010-002 .

[23] A. D. Martin et al., "Parton distributions for the LHC", Eur. Phys. J. C63 (2009) 189.

Jet and Multijet Results from CMS

Konstantinos Kousouris
Fermi National Accelerator Laboratory, Batavia IL, 60510, USA
E-mail: `kkousour@fnal.gov`

Abstract. QCD measurements with jets are presented, from proton-proton collisions at a centre-of-mass energy of 7 TeV at the CERN LHC. The data were collected with the CMS detector during the 2010 data-taking period, and correspond up to an integrated luminosity of $36\,\mathrm{pb}^{-1}$. The measured inclusive-jet and dijet production cross sections are compared to perturbative QCD predictions at next-to-leading-order, and are found to be in good agreement. Observables sensitive to multijet production, such as the hadronic event shapes, the dijet azimuthal decorrelations, and the ratio of the 3-jet to 2-jet production cross section, are compared to the predictions of various QCD Monte-Carlo generators.

1. Introduction

In Quantum Chromodynamics (QCD), events with two or more high transverse-momentum jets arise in proton-proton collisions from parton-parton scattering, where the outgoing scattered partons manifest themselves as hadronic jets. The inclusive jet and dijet production cross sections are benchmarks of the standard model (SM), and confront the perturbative QCD predictions at the smallest distance scales ever probed. The cross-section measurements can be used to constrain the parton momentum distributions (PDF) in the proton, and are sensitive to the strong coupling constant. Significant deviations from the predictions of the inclusive jet and dijet cross sections at high transverse momentum or dijet invariant mass, could also be an indication of new phenomena beyond the SM. Other observables, such as the dijet azimuthal decorrelations, the hadronic event shapes, and the ratio of the 3-jet to 2-jet production cross section, are sensitive to multijet production and can be used to differentiate between the various QCD Monte-Carlo generators.

In this paper, various QCD measurements with jets are reported, using proton-proton collisions at a centre-of-mass energy of 7 TeV at the CERN Large Hadron Collider (LHC). The data were collected with the Compact Muon Solenoid (CMS) detector during the 2010 run, and correspond to an integrated luminosity of up to $36\,\mathrm{pb}^{-1}$. The paper is organized as follows: the CMS detector is described briefly in Section 2, and the jet reconstruction approach is described in Section 3. The inclusive jet and dijet cross section measurements are reported in Section 4, while the multijet-sensitive measurements are discussed in Section 5. Section 6 summarizes the results.

2. The CMS Detector

The CMS coordinate system has its origin at the centre of the detector, with the z-axis pointing along the direction of the counterclockwise beam. The azimuthal angle is denoted as ϕ, the polar angle as θ, and the pseudorapidity is defined as $\eta = \ \ln[\tan(\theta/2)]$. The central feature of the CMS detector is a superconducting solenoid, of 6 m internal diameter, that produces an axial magnetic field of 3.8 T. Within the field volume are the silicon pixel and strip tracker, a lead-tungstate crystal electromagnetic calorimeter (ECAL) and a brass/scintillator hadronic calorimeter. Outside the field volume, in the forward region ($3 < |\eta| < 5$), is an iron/quartz-fiber hadronic calorimeter. Muons are measured in gas detectors embedded in the steel return yoke outside the solenoid, in the pseudorapidity range $|\eta| < 2.4$. A detailed description of the CMS experiment can be found in Ref. [1].

3. Jet Reconstruction

Jets are reconstructed using the anti-k_T clustering algorithm [2] with size parameters $R = 0.5$ and $R = 0.7$. The clustering is performed using four-momentum summation. The rapidity y and the transverse momentum p_T of a jet with energy E and momentum $\vec{p} = (p_x, p_y, p_z)$ are defined as $y = \frac{1}{2}\ln\left(\frac{E+p_z}{E-p_z}\right)$ and $p_T = \sqrt{p_x^2 + p_y^2}$, respectively. The inputs to the jet clustering algorithm are the four-momentum vectors of the reconstructed particles. Each such particle is reconstructed with the particle-flow technique [3] which combines the information from several subdetectors. The resulting jets require an additional energy correction to take into account the non-linear and non-uniform response of the CMS calorimetric system to the neutral-hadron component of the jet (the momentum of charged hadrons and photons is measured accurately by the tracker and the ECAL, respectively). The jet-energy corrections are derived using simulated events, generated by PYTHIA6.4.22 (PYTHIA6) [4] and processed through the CMS detector simulation based on GEANT4 [5], and *in situ* measurements with dijet and photon+jet events [6]. An offset correction is also applied to take into account the extra energy clustered in jets due to additional proton-proton interactions within the same bunch crossing (pile-up). The jet-energy correction depends on the η and p_T of the jet, and is applied as a multiplicative factor to the jet four-momentum vector. The multiplicative factor is in general smaller than 1.2, and approximately uniform in η. For a jet $p_T = 100\,\mathrm{GeV}$ the typical factor is 1.1, decreasing towards 1.0 with increasing p_T. The dijet invariant mass is calculated from the corrected four-momentum vectors of the two jets with the highest p_T (leading jets): $M_{JJ} = \sqrt{(E_1 + E_2)^2 - (\vec{p}_1 + \vec{p}_2)^2}$.

4. Cross-Section Measurements

The data samples used for this measurement were collected with single-jet high level triggers (HLT) [7] which required at least one jet in the event to satisfy the condition $p_T > 30$, 50, 70, 100 and 140 GeV, respectively, in uncorrected jet transverse momentum. The online jet reconstruction used only calorimetric information and the resulting HLT jets had typically much lower energy response than the offline particle-flow jets. For the construction of the inclusive-jet and dijet spectra, each bin is populated by events collected only with the fully efficient trigger with the highest threshold.

The double-differential inclusive jet cross section [8], as a function of the jet p_T and rapidity, is defined as:

$$\frac{d^2\sigma}{dp_T d|y|} = \frac{C_{uns}}{\epsilon\mathcal{L}} \frac{N_{jets}}{\Delta p_T \Delta|y|}, \tag{1}$$

where N_{jets} is the number of jets, Δp_T and $\Delta|y|$ are the widths of the p_T and rapidity bins, respectively, ϵ is the event selection efficiency, $\mathcal{L}$ is the integrated luminosity and C_{uns} is a correction factor for the detector smearing effects. The latter ranges from 10% in the central rapidity bins, to 50% in the outermost bin.

The double-differential dijet cross section [9] is measured as a function of the dijet invariant mass M_{JJ} and $|y|_{max} = \max(|y_1|, |y_2|)$, and is expressed similar to eqn.1:

$$\frac{d^2\sigma}{dM_{JJ}d|y|_{max}} = \frac{C_{uns}}{\epsilon\mathcal{L}} \frac{N_{evt}}{\Delta M_{JJ}\Delta|y|_{max}}, \tag{2}$$

where N_{evt} is the number of dijet events. The unsmearing correction C_{unsm} for the dijet cross section is smaller than 10% in the entire phase space of the measurement. In order to capture most of the gluon radiation and improve the dijet mass resolution, the measurement is performed using the larger jet size of $R = 0.7$, as opposed to the other measurements presented in this paper, which use the default $R = 0.5$ CMS size.

Figure 1 shows the inclusive-jet and dijet production cross sections, compared to the theory predictions. The latter consists of a perturbative QCD (pQCD) calculation at next-to-leading-order (NLO) and a non-perturbative correction to account for the multiple-parton interactions (MPI) and the hadronization effects. For the pQCD calculation, the PDF4LHC prescription [10] was used. The non-perturbative correction was calculated from various Monte-Carlo generators and ranges from 30% at low jet p_T (dijet mass) to a few percent at higher values. A more detailed comparison to the theory is shown in Fig. 2.

The experimental systematic uncertainty is dominated by the jet-energy scale uncertainty, which varies between 3-5% and results in a 20-50% cross-section uncertainty. The luminosity uncertainty is estimated to be 4% and propagates directly to the cross section. The theoretical uncertainty is dominated by the non-perturbative correction and the PDF uncertainty at low and high jet p_T (dijet mass) respectively.

The inclusive-jet cross section spans a p_T range from 18 GeV to 1.1 TeV and up to $|y| = 3$ in jet rapidity. The dijet cross section spans an M_{JJ} range from 156 GeV to 3.5 TeV. Both cross-section measurements are found to be in good agreement with the theory predictions, in the entire phase space.

5. Multijet Measurements

The understanding of the QCD multijet production is critical for searches of physics beyond the SM, in hadronic final states. A typical example is Supersymmetry, where QCD is a dominant background. Observables which are sensitive to the QCD multijet production can be used to tune the various Monte-Carlo generators. In addition, they can be used to measure the strong coupling constant α_S.

The dijet azimuthal decorrelation is a simple observable that probes indirectly the multijet production. The normalized dijet cross section $\frac{1}{\sigma_{dijet}}\frac{dN}{d\Delta\phi}$, as a function of the azimuthal separation $\Delta\phi$ between the two leading jets, is measured in bins of the leading jet p_T and provides an indirect handle on the multijet production: the $\Delta\phi$ values close to π denote the back-to-back dijet production, while values close to $\pi/2$ denote the 3-jet or 4-jet region. This observable is free of many experimental uncertainties (e.g. jet-energy scale, luminosity), and the measurement was performed in CMS using the first 2.9 pb^{-1} of data [11].

Figure 3 shows the dijet azimuthal decorrelation compared to the pQCD predictions at leading-order (LO) and NLO. It also shows the measurement compared to the prediction from various Monte-Carlo generators. The LO prediction, as expected, fails to describe the dijet azimuthal decorrelation, while the NLO prediction improves the agreement in the range $\Delta\phi \sim 2\pi/3$-π, which corresponds roughly to the 2-jet and 3-jet production. The Monte-Carlo generators are able to describe the measurement reasonably well in the entire range.

The hadronic event-shape variables probe directly the QCD multijet production. In particular, two event-shape variables have been studied: the central transverse thrust $\tau_{\perp,\mathcal{C}}$ and the central thrust minor $T_{m,\mathcal{C}}$. The two variables probe different QCD radiative processes and are mostly sensitive to the modeling of two- and three-jet topologies. The term *central* ($\mathcal{C}$) indicates that the input to the calculation of these quantities are jets in the central region of

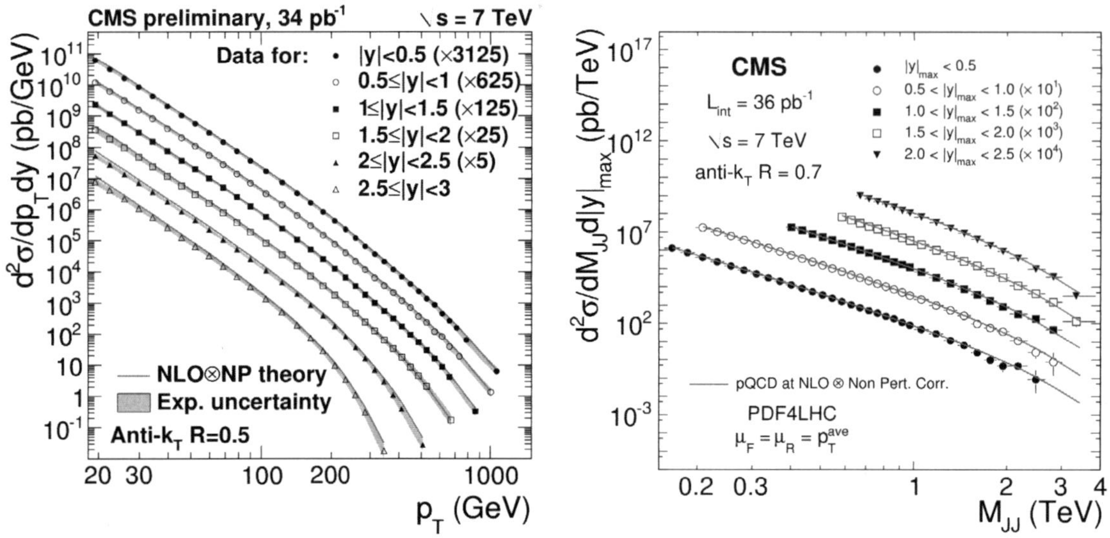

Figure 1. Left: differential inclusive-jet production cross section, as a function of jet p_T in various rapidity bins. Right: differential dijet production cross section, as a function of the dijet invariant mass, in various rapidity bins. The cross sections are scaled by the factors shown in the figures. The points represent the measured cross sections, corrected for detector effects, while the curves represent the pQCD predictions at next-to-leading-order, corrected for non-perturbative effects.

the detector ($|\eta| < 1.3$), where sub-leading contributions in the calculation of the event-shape variables are less significant, and systematic uncertainties on the jet reconstruction are smaller. The central transverse thrust is defined as:

$$\tau_{\perp,\mathcal{C}} \equiv 1 - \max_{\hat{n}_T} \frac{\sum_i |\vec{p}_{\perp,i} \cdot \hat{n}_T|}{\sum_i p_{\perp,i}}, \tag{3}$$

where $p_{\perp,i}$ is the transverse momentum of selected jet i. The axis $\hat{n}_T$ which maximizes the sum, and thus minimizes $\tau_{\perp,\mathcal{C}}$, is called the transverse thrust axis $\hat{n}_{T,\mathcal{C}}$. The central transverse thrust is a measure of the radiation along the transverse thrust axis. The central thrust minor is a measure of the radiation out of the plane defined by $\hat{n}_{T,\mathcal{C}}$ and the beam axis. It is defined as:

$$T_{m,\mathcal{C}} \equiv \frac{\sum_i |\vec{p}_{\perp,i} \times \hat{n}_{T,\mathcal{C}}|}{\sum_i p_{\perp,i}}. \tag{4}$$

Two-jet events that are well balanced have low values of these two variables, while isotropic multijet events have high values. The measurement is free of many experimental uncertainties, and has been performed using $3.2\,\text{pb}^{-1}$ of data [12].

Figure 4 shows the distributions of the logarithms of the central transverse thrust and the central thrust minor variables, for events where the leading-jet $p_T > 200\,\text{GeV}$. The measurements are compared to the predictions of various Monte-Carlo generators. For the given generator tunes considered here, PYTHIA6 and HERWIG++ [13] describe well the measurement in both the 2-jet and 3-jet region. The other generators, namely (PYTHIA8 [14], ALPGEN [15], MADGRAPH [16]) are less successful in describing the data.

Another direct probe of the multijet production, is the ratio of the 3-jet over 2-jet production cross section. The measurement has been performed using the entire 2010 CMS dataset [17],

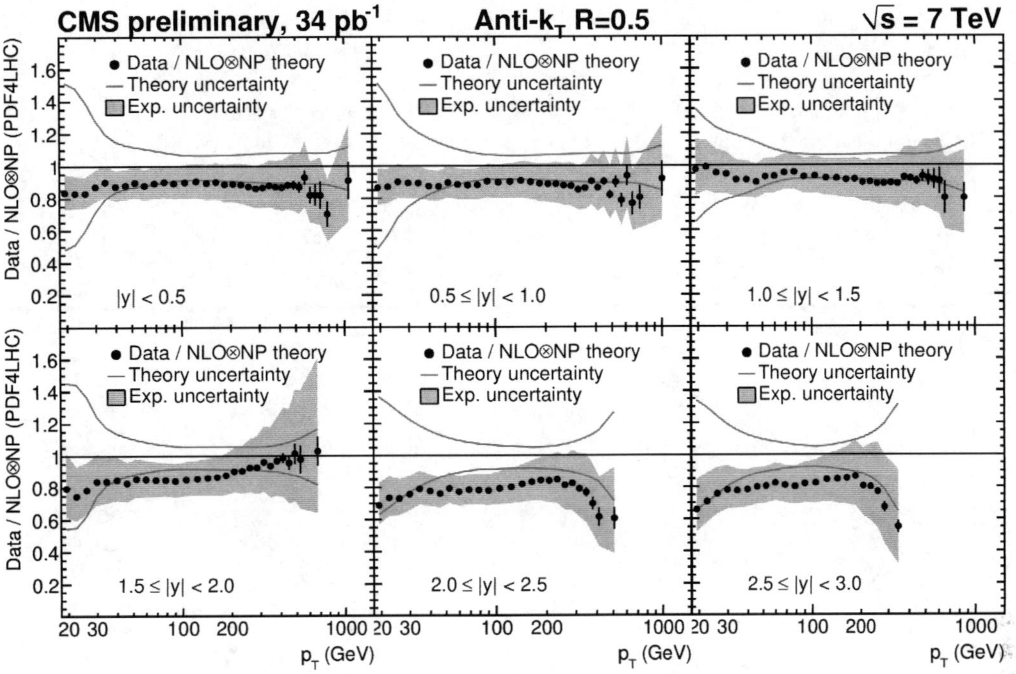

Figure 2. Ratio of the measured differential inclusive-jet cross section, over the theory prediction, in different rapidity bins. The shaded band represents the total experimental systematic uncertainty and is drawn around the data points. The curves represent the total theoretical uncertainty and are drawn around the horizontal axis.

and expressed as a function of H_T. The latter is the scalar sum of the jet transverse momenta: $H_T = \sum_j p_{T,j}$. Many experimental systematic uncertainties are cancelled in the ratio, leading to a robust measurement, covering the H_T range from 200 GeV to 2.5 TeV. The 3-jet over 2-jet ratio, is in general sensitive to the strong coupling constant α_S, but the format examined here (as a function of H_T) is more suitable for studies of the soft jet production. Thus, the measurement is compared to the predictions from Monte-Carlo generators, after correcting for detector smearing effects. Figure 5 shows the measured 3-jet to 2-jet cross-section ratio, compared to different models. The MADGRAPH generator, interfaced with PYTHIA6 for the MPI and hadronization, gives the best description of the data.

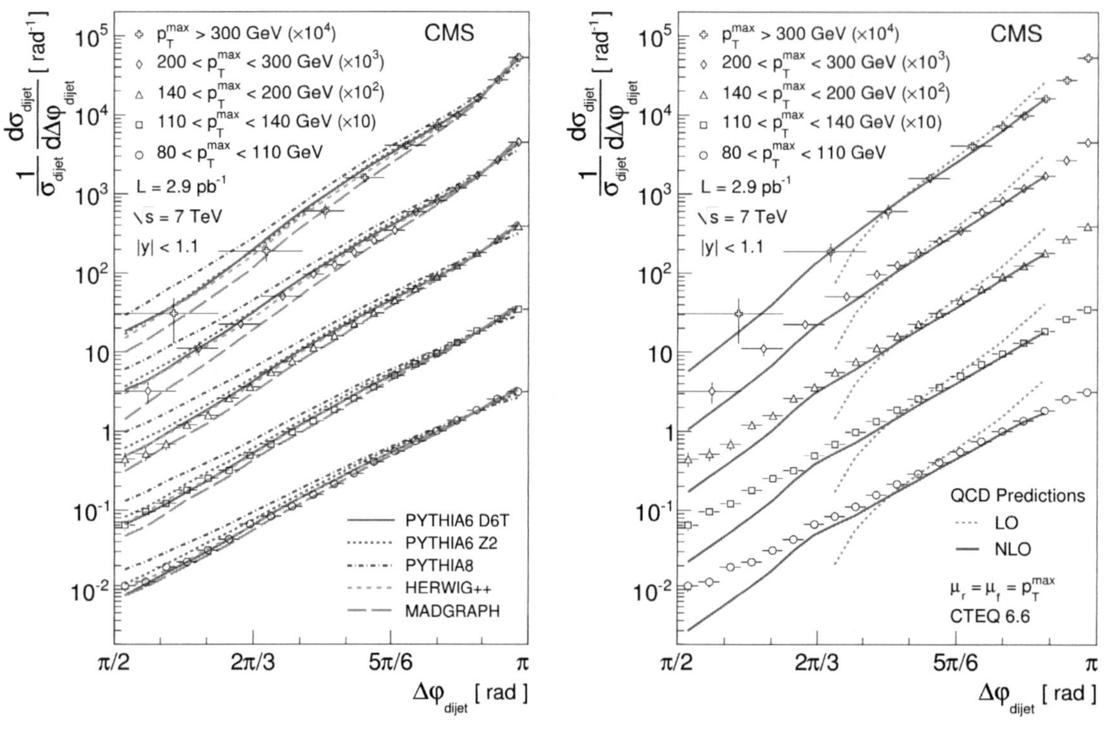

Figure 3. Normalized $\Delta\phi_{dijet}$ distributions in several $p_{T,max}$ regions, scaled by the multiplicative factors given in the figure for easier presentation. Left: the curves represent predictions from various Monte-Carlo generators. Right: the curves represent predictions from LO (dotted line) and NLO pQCD (solid line). Non-perturbative corrections have been applied to the predictions. The error bars on the data points include statistical and systematic uncertainties.

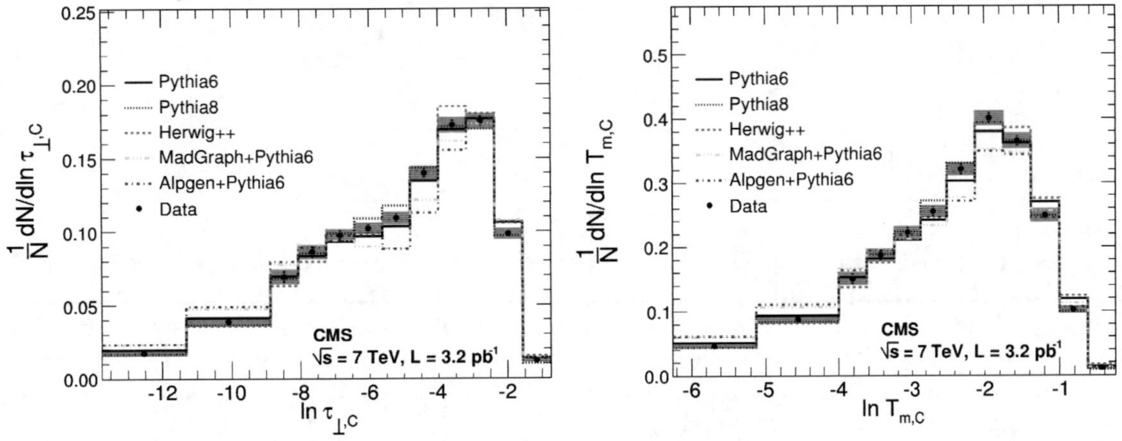

Figure 4. Distributions of the logarithm of the central transverse thrust (left) and central thrust minor (right) for events with a leading jet $p_T > 200\,\text{GeV}$, from data and from five MC simulations. The error bars on the data points represent the statistical uncertainty on the data, and the shaded bands represent the sum of statistical and systematic errors.

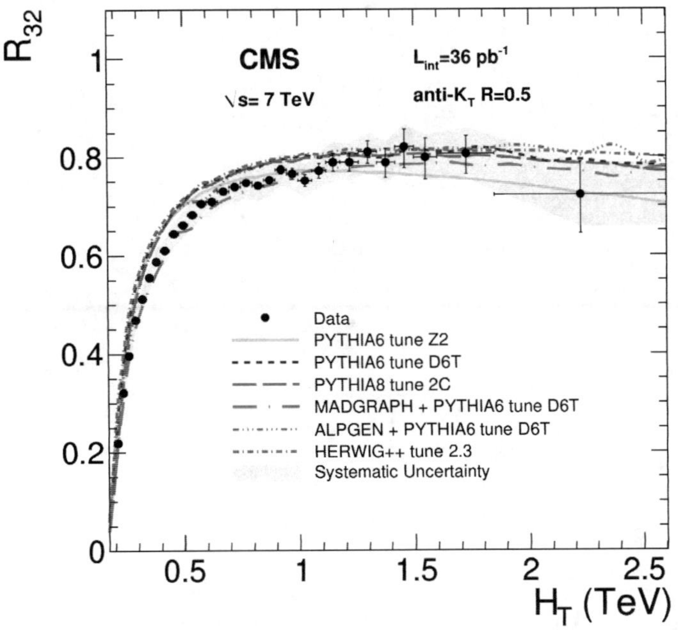

Figure 5. The measured 3-jet over 2-jet cross-section ratio (filled circles) as a function of H_T, and the predictions of Monte-Carlo generators (curves). Error bars represent statistical uncertainties. The shaded area indicates the size of the combined systematic uncertainty.

6. Summary

CMS has performed various QCD studies with jets, using proton-proton collisions at $\sqrt{s} = 7\,\text{TeV}$, from the 2010 LHC run, corresponding to integrated luminosity up to $36\,\text{pb}^{-1}$. The inclusive-jet and dijet production cross sections have been measured in a previously unexplored kinematic region, reaching jet-p_T values beyond 1 TeV and dijet invariant masses up to 3.5 TeV. Both measurements are in good agreement with the pQCD predictions at NLO. The dijet azimuthal decorrelations and the hadronic event-shape variables have been also measured, using approximately $3\,\text{pb}^{-1}$ of data and compared to Monte-Carlo generator predictions. Both measurements are well reproduced by the PYTHIA6 and HERWIG++ generators. Finally, the 3-jet over 2-jet production ratio as a function of H_T is best reproduced by MADGRAPH, interfaced with PYTHIA6.

References

[1] CMS Collaboration 2008 *JINST* **3** S08004
[2] Cacciari M, Salam G P and Soyez G 2008 *JHEP* **0804** 063
[3] CMS Collaboration 2009 *CMS PAS* **PFT-09-001**
[4] Sjöstrand T, Mrenna S and Skands P 2006 *JHEP* **05** 026
[5] Agostinelli S et al 2003 *Nucl. Inst. Meth.*A **506** 250
[6] CMS Collaboration 2010 *CMS PAS* **JME-10-010**
[7] CMS Collaboration 2006 *Eur. Phys. J.* C **46** 605
[8] CMS Collaboration 2011 *CMS PAPER* **QCD-10-011**
[9] CMS Collaboration 2011 *Phys. Lett.* B **700** 187-206
[10] Alekhin S et al 2011 *arXiv:1101.0536v1*
[11] CMS Collaboration 2011 *Phys. Rev. Lett.* **106** 122003
[12] CMS Collaboration 2011 *Phys. Lett.* B **699** 48-67
[13] Bähr M et al 2008 *Eur. Phys. J.* C **58** 639
[14] Sjöstrand T, Mrenna S and Skands P 2008 *Comput. Phys. Commun.* **178** 852
[15] Mangano M L et al 2003 *JHEP* **07** 001
[16] Alwall J et al 2007 *JHEP* **09** 028
[17] CMS Collaboration 2011 *CMS PAPER* **QCD-10-012**

Searches for new physics with jets in ATLAS

Belen Salvachua, for the ATLAS Collaboration
High Energy Physics Division
Argonne National Laboratory
Argonne, IL 60439 USA

E-mail: belen.salvachua@cern.ch

Abstract. We present the latest results of searches for new physics beyond the Standard Model with jets in the final state using the ATLAS detector. These analyses are performed with the full LHC 2010 data from proton-proton collisions at a center-of-mass energy of 7 TeV. The results are based on an integrated luminosity of 33 to 37 pb^{-1} depending on the analysis. No significant discrepancy is found with the expected Standard Model predictions. New limits on various models are set beyond the reach of previous experiments.

1. Introduction

The Standard Model (SM) describes most of the current high energy physics data well. It accommodates three families of leptons and quarks that couple via the exchange of four bosons. The mass of these particles is generated via the spontaneous symmetry breaking mechanism at the cost of predicting a new particle, the Higgs boson, which has not been observed yet. Even if the Higgs boson is observed, there are still many open questions not explained within the Standard Model such as the apparent asymmetry of matter-anti-matter in the visible universe or what is the mechanism that generates the mass of fermions. Since 2009, the Large Hadron Collider (LHC) has been producing proton-proton collisions to investigate some of these questions. The data recorded by the ATLAS experiment during 2010 includes 45 pb^{-1} at a center-of-mass energy of 7 TeV. Therefore, we are now accessing a new energy regime where new physics could potentially appear.

In this note, we will present the search for new physics in final states containing jets. The analysis try to be model-independent when possible and the results are then interpreted within different models. Section 3 shows the results on the dijet final state where we are looking for new resonances and new interactions by analyzing the dijet mass spectrum and angular distributions. Section 4 shows the search for a heavy fourth generation quark (Q_4) in final states with two W bosons and jets. And finally, Section 5 shows a search for 1st and 2nd generation leptoquark (LQ) pair production in leptons-plus-jets signatures.

2. Detector description

The ATLAS detector is described in detail elsewhere [1]. The beam-line is surrounded by a tracking detector that uses silicon pixel, silicon strip, and straw tubes and is embedded in a 2 T magnetic field. The tracking system covers the pseudorapidity [2] range $|\eta| < 2.5$. It is surrounded by electromagnetic and hadronic calorimeters covering $|\eta| < 3.2$, which are complemented by a forward hadronic calorimeter covering $3.1 < |\eta| < 4.9$. The luminosity calibration has been determined during dedicated van der Meer beam scans to a precision of 3.4% [3, 4].

3. Dijet signatures

The aim of this analysis is to search for massive objects and new interactions by studying the dijet invariant mass and angular distributions. Previous ATLAS results used 3.1 pb^{-1} [5] while the results presented here use an order of magnitude more of data, 36 pb^{-1} integrated luminosity. The dijet inclusive cross-section has been measured in ATLAS up to dijet masses of 4.1 TeV [6] testing the Standard Model in a new unexplored regime.

The dijet invariant mass spectrum is sensitive to new phenomena: potential new physics could show up as a local excess of events in the dijet mass spectrum. This analysis starts with a basic event pre-selection that requires a primary vertex with more than 4 charged tracks and at least 2 jets with $p_T^1 > 60$ GeV and $p_T^2 > 30$ GeV. For the dijet analysis [7], we use all 2010 data that pass a first-level inclusive jet trigger which has an efficiency larger than 99% for jet pairs with $m_{jj} > 500$ GeV.

For the resonance search additional event selection cuts are applied. Only events with a leading jet with transverse momentum larger than 150 GeV, dijet invariant mass, m_{jj}, larger than 500 GeV, $|\eta_j| < 2.5$ and $\Delta\eta_{jj} < 1.3$ are used.

This analysis looks for resonances in the dijet invariant mass with a data-driven method that models the rapidly falling smooth distribution of the SM dijet background shape using the formula:

$$f(x) = p_1(1-x)^{p_2} x^{p_3+p_4 \ln x} \quad \text{with} \quad x = m_{jj}/\sqrt{s} \quad , \tag{1}$$

where p_i are the parameters of the fit, $\sqrt{s}$ is the center-of-mass energy and m_{jj} the invariant mass of the two leading jets. This formula has been empirically shown to model the shape of SM dijet background [8, 9]. After the fit, a χ^2 test is performed, resulting in a p-value of 0.88 which is indeed in good agreement with the QCD expected shape. An additional test is performed using a BumpHunter algorithm [10]. This algorithm searches for a signal window with an excess of events. The largest discrepancy is found in the region 995 GeV $< m_{jj} <$ 1253 GeV with a p-value of 0.39 which is still very good agreement between the data and the fit to a smooth rapidly falling distribution. Figure 1 shows the observed dijet mass distributions fitted using Eq. 1. Overlaid to the distribution are the predicted signals for excited quarks ($q^\star$) with masses of 1000, 1700 and 2500 GeV normalized to 36 pb^{-1}. The bin-by-bin significance of the data-background difference is shown in the lower panel.

We do not find any discrepancy with the Standard Model prediction and are able to set new exclusion limits in several models. Using Bayesian credibility intervals, we exclude the production of an excited quark $q^\star$ with masses below 2.15 TeV (see Figure 2) and the production of an axigluon with masses below 2.10 TeV. We also set limits on "low multiplicity" quantum black hole models (QBH) [6, 11]. This model produces threshold effects in m_{jj} with long tails to higher dijet masses that compete with the SM dijet background. Since, the cross-section is very large just above the threshold it is possible to extract limits given the resulting resonance-like signal shape. The resulting limits are shown in Table 1.

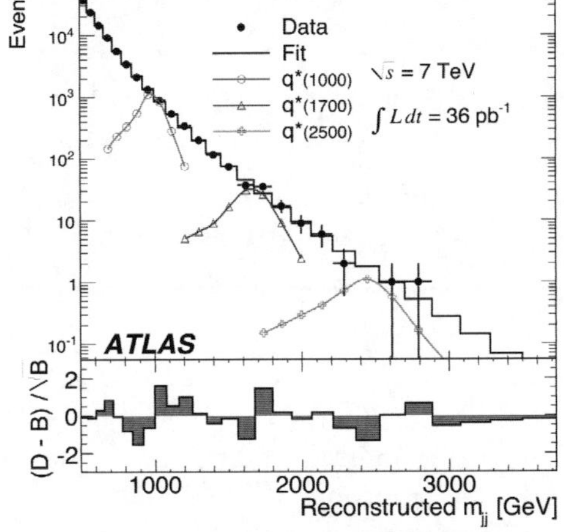

Figure 1: Observed dijet mass distribution (filled dots) and the fitted function (solid line). The bin-by-bin significance of the data-background difference is shown in the lower panel.

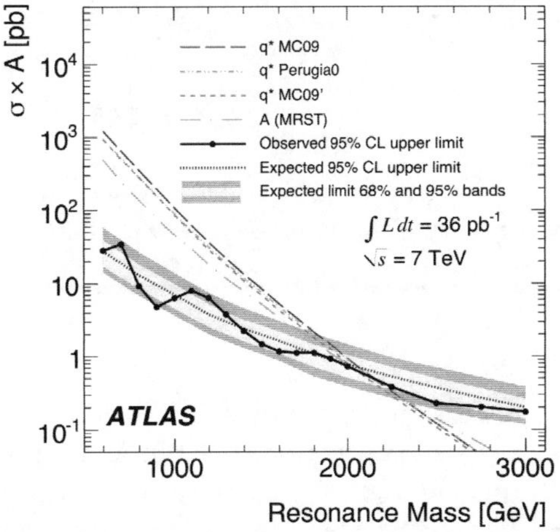

Figure 2: The 95% C.L. upper limits on the cross-section times acceptance for a resonance decaying to dijets compared to an axigluon model and to a $q^\star$ model.

n Extra Dimensions	Observed Limit [TeV]	Expected Limit [TeV]
2	3.20	3.18
3	3.38	3.35
4	3.51	3.48
5	3.60	3.58
6	3.67	3.64
7	3.73	3.71

Table 1: The 95% C.L. observed and expected lower limits on the allowed quantum gravity scale, M_D, for various numbers of extra dimensions.

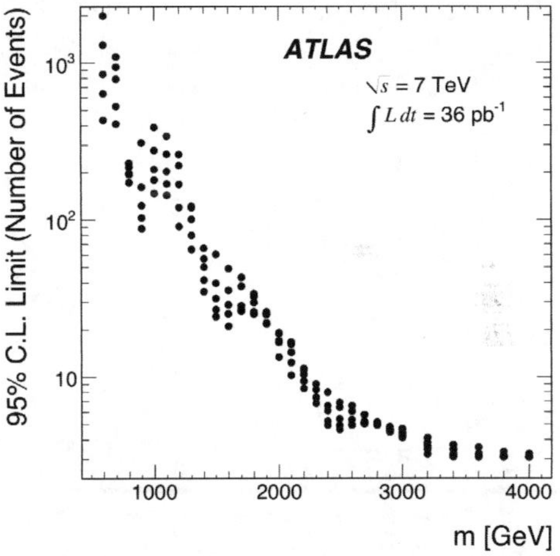

Figure 3: Production cross-section limits as a function of dijet mass, for several values of the resonance width.

As a novelty, in this analysis, we also set up model-independent limits by using a simplified Gaussian method. Our signal template is a Gaussian with mean (μ) that varies between 600 GeV and 4000 GeV and sigma (σ) between 3% and 15%. We set production cross-section limits as a function of the dijet mass to facilitate comparisons with other hypotheses. Figure 3 shows the 95% C.L. limits as a function of the resonance mass expressed in terms of number of events observed after applying all event selection criteria. Points in the same mass bin correspond to different resonance widths.

For the analysis on the angular distributions the starting point is the sample obtained with the pre-selection described above. We define 2 variables:

$$y_B = \frac{1}{2}(y_1 + y_2) \quad \text{and} \quad y^\star = \frac{1}{2}(y_1 - y_2) \quad , \tag{2}$$

where y_i is the rapidity of jet i. We then select events with $|y_B| < 1.1$, $|y^\star| < 1.7$ and $|y_{1,2}| < 2.8$. The dijet invariant mass is required to be $m_{jj} > 500$ GeV. Figure 4 shows the distribution of the angular variable χ defined as

$$\chi = \exp(|y_1 - y_2|) \quad , \tag{3}$$

for several dijet mass bins. This variable is shown to be relatively flat for the expected QCD background. In the Figure we compare the QCD prediction with a quantum black hole scenario with quantum gravity mass scale of $M_{\rm D} = 3$ TeV and six extra dimensions.

Figure 5 shows the fraction of dijets produced centrally versus the total number of observed dijets for a specified dijet mass range:

$$F_\chi([m_{jj}^{min} + m_{jj}^{max}]/2) = \frac{N_{\rm events}(|y^\star| < 0.6, m_{jj}^{min}, m_{jj}^{max})}{N_{\rm events}(|y^\star| < 1.7, m_{jj}^{min}, m_{jj}^{max})} \quad . \tag{4}$$

This variable is sensitive to mass-dependent changes. In the Figure, the observed data $F_\chi(m_{jj})$ are compared with QCD predictions and with the predicted distribution for a contact interaction model with the compositeness scale $\Lambda = 5.0$ TeV. The data are compatible with the background-only hypothesis and exclusion limits are set from this analysis, see Ref. [7]. In particular, we exclude contact interactions forΛ < 9.5 TeV at 95% C.L.

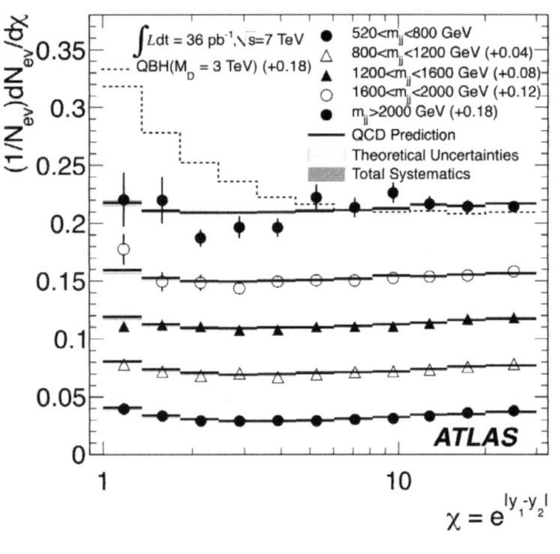

Figure 4: The χ distributions for several dijet mass bins shifted up by the value shown in the legend.

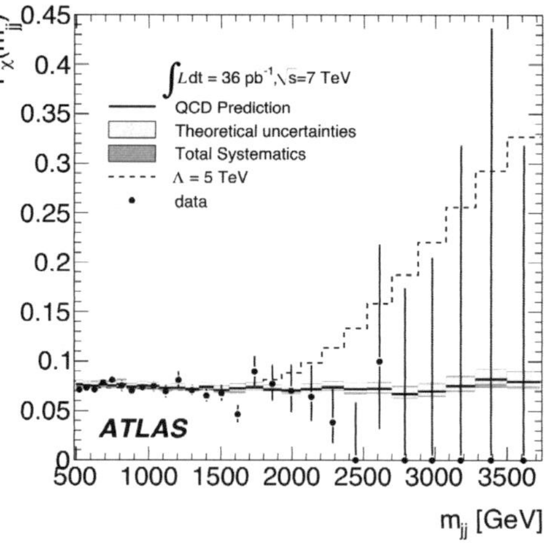

Figure 5: The $\mathrm{F}_\chi(\mathrm{m}_{jj})$ function versus dijet invariant mass.

4. WWjj signatures

In this note, we also show a search for pair production of fourth generation quarks, Q_4, decaying $Q_4 \to Wq$ where the quark type q could be u, d, c, s or b. We assume a u_4 model for the analysis but the results are directly applicable to more exotic quark models (*i.e.* quarks with charges $-1/3, -4/3$ decaying into Wq, q being a light quark) [12]. We analyze the case of both W bosons decaying into $l\nu$; therefore the final state will contain two oppositely charged leptons, jets and missing transverse energy from the undetected neutrinos.

The initial event selection suppresses most of the background contribution from $Z/\gamma^\star$ production; this selection requires:

- At least 2 jets with $p_{\mathrm{T}} > 20$ GeV and $|\eta| < 2.5$.
- For same flavor events (ee and $\mu\mu$):
  - Missing transverse energy $E_{\mathrm{T}}^{\mathrm{miss}} > 40$ GeV.
  - Dilepton invariant mass, $m_{ll} > 15$ GeV, as the Monte-Carlo does not include contribution from lower invariant masses.
  - Dilepton invariant mass must fall outside a m_Z window, $m_{ll} < 81$ GeV or $m_{ll} > 101$ GeV.
- For different flavor events ($e\mu$):
  - Scalar sum of transverse energy from leptons and jets $H_{\mathrm{T}}^{jet,leptons} > 130$ GeV.

Figure 6 shows the missing transverse energy for the data (points) and the expected signal and backgrounds for the 3 leptonic channels ee, $\mu\mu$ and $e\mu$ after this baseline selection. Afterwards, a more refined selection uses the mass reconstruction to suppress the dominant top quark pair background.

We can reconstruct both neutrinos by assuming that the two neutrinos are the only contribution to the $E_{\mathrm{T}}^{\mathrm{miss}}$ and that they are approximately collinear with the leptons. In order to do this, we vary $|\Delta\eta(\nu l)|$ and $|\Delta\phi(\nu l)|$ independently between 0 and 1 and reconstruct the invariant masses of the two $l\nu q$ systems. We choose the values of $|\Delta\eta(\nu l)|$, $|\Delta\phi(\nu l)|$ and the jet assignment that minimize the difference of the two reconstructed collinear masses ($M_{\mathrm{collinear}}$) in the event. Figure 7 shows the distribution of the sum of transverse energy versus collinear mass for background (left) and signal of a heavy quark with mass $m_{Q_4} = 350$ GeV for the sum of the ee, $\mu\mu$ and $e\mu$ channels. These distributions indicate that it is possible to remove a significant fraction of the background events by applying a triangular cut dependent on $H_{\mathrm{T}} = H_{\mathrm{T}}^{jet,leptons} + E_{\mathrm{T}}^{\mathrm{miss}}$ and $M_{\mathrm{collinear}}$. Table 2 shows the final selection cuts for each Q_4 mass.

Table 2: List of final selection cuts for each Q_4 mass.

Q_4 Mass [GeV]	Final selection
250	$H_{\mathrm{T}} > 500 - 0.7 \times M_{\mathrm{collinear}}$
300	$H_{\mathrm{T}} > 600 - 0.5 \times M_{\mathrm{collinear}}$
350	$H_{\mathrm{T}} > 600 - 0.2 \times M_{\mathrm{collinear}}$
400	$H_{\mathrm{T}} > 700 - 0.3 \times M_{\mathrm{collinear}}$

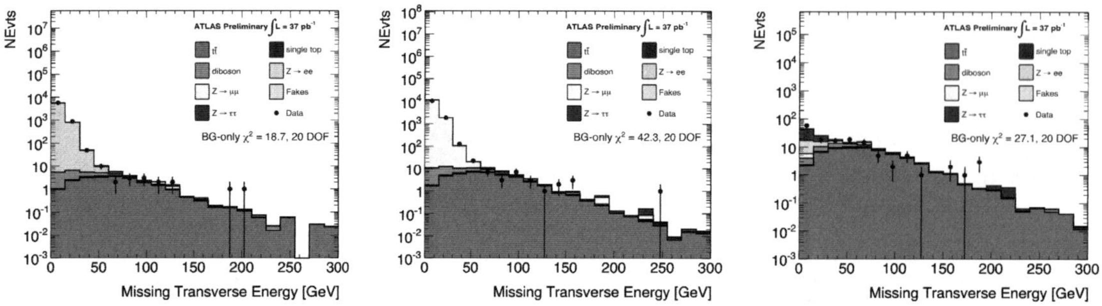

Figure 6: Missing transverse energy, $E_{\mathrm{T}}^{\mathrm{miss}}$ after dilepton selection in ee, $\mu\mu$ and $e\mu$ channels.

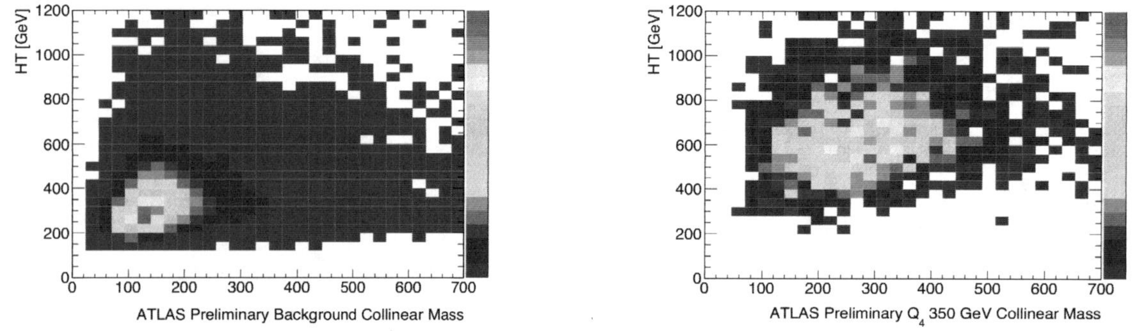

Figure 7: Sum of transverse energy H_{T} versus $M_{\mathrm{collinear}}$ for background (*left*) and for a heavy quark Q_4 with mass $m_{Q_4} = 350$ GeV (*right*) for the sum of ee, $\mu\mu$ and $e\mu$ channels.

Figure 8 shows the collinear mass distribution after the final selection cuts. The data are consistent with the expectation from the Standard Model and we set a lower limit on the Q_4 mass of $m_{Q_4} > 270$ GeV at 95% C.L., see Figure 9. These limits are applicable to u_4-type quarks as well as other exotic quark models.

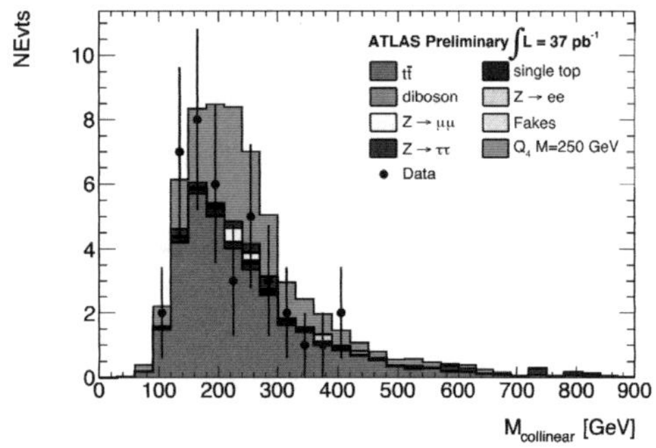

Figure 8: Reconstructed collinear mass ($M_{\mathrm{collinear}}$) for the sum of ee, $\mu\mu$ and $e\mu$ channels after applying the final cuts for expected signal of $m_{Q_4} = 250$ GeV.

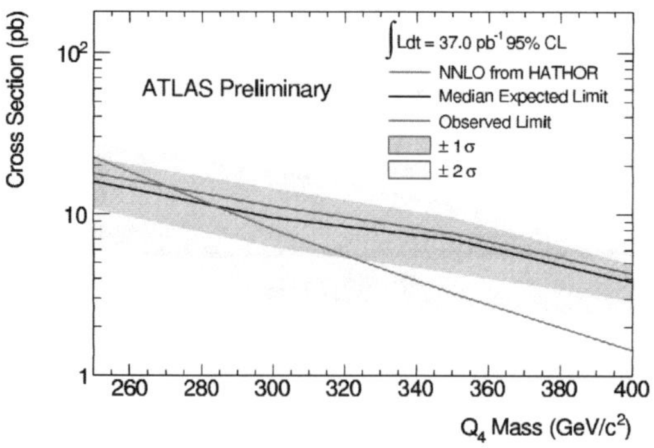

Figure 9: 95% confidence level observed and median expected cross-section upper limits compared to the theoretical prediction.

5. Leptons plus jets signatures

We have also investigated the production of leptoquark pairs decaying into leptons plus jets. Leptoquarks are particles that carry both lepton and baryon number; these particles are predicted by theories such as substructure theories, grand unification and extended technicolor. The cross-sections times branching ratios for the leptoquark-mediated processes $pp \to lljj$ and $l\nu jj$ can be written as $\sigma_{\rm LQ} \times \beta^2$ and $\sigma_{\rm LQ} \times 2\beta(1-\beta)$ respectively, where β is the branching fraction for a single leptoquark to decay into a charged lepton and a quark. The production rates also depend strongly on the leptoquark mass $M_{\rm LQ}$ and therefore the final production limits will be shown as a function of these two parameters $M_{\rm LQ}$ and β.

The main backgrounds for the channel $lljj$ are $Z+$jets and $t\bar{t}$ production and for the channel $l\nu jj$, $W+$jets and $t\bar{t}$ production. For leptoquark pair production we expect a signal peak in the invariant mass of jet-lepton (and jet-neutrino) pair together with large $S_{\rm T}^l$, where $S_{\rm T}^l$ is the scalar sum of the transverse momentum of the two charged leptons and jets; in the case of the $l\nu jj$ channel the missing transverse energy is used on the calculation. Figure 10 shows the $S_{\rm T}^l$ distribution for the $eejj$ final state after all selections. Figure 11 shows the invariant mass (lepton-jet) distribution for the $e\nu jj$ channel after all selections. In both cases, the data are indicated by the points and the Standard Model backgrounds are shown with cumulative distributions. The expected LQ signals for various masses are also shown.

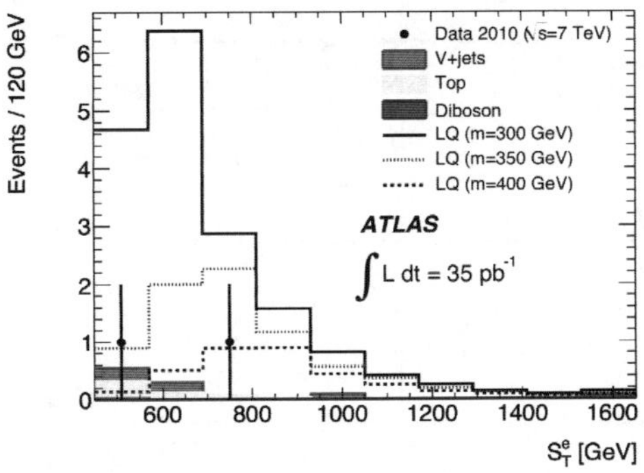

Figure 10: $S_{\rm T}^e$ distribution for the $eejj$ final state signal region.

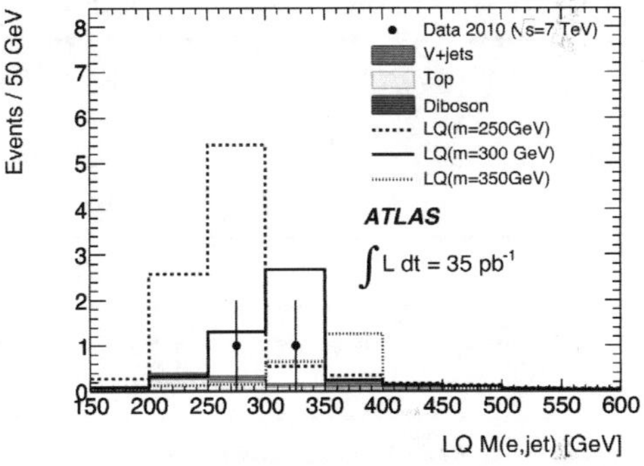

Figure 11: $M_{\rm LQ}$ distribution for the $e\nu jj$ final state signal region.

The data match the background-only prediction. Figures 12 and 13 show the 95% C.L. exclusion region obtained from the combination of the two electron channels (left) and the two muon channels (right) in the β versus $M_{\rm LQ}$ plane. The gray area indicates the D0 exclusion limit and the thick dotted line the CMS exclusion. The dotted and dotted-dashed lines show the individual limits for the $lljj$ and the $lvjj$ channels, respectively. The combined expected limit is indicated by the dashed line. The combined observed limit is indicated by the solid line. We set 95% C.L. observed lower limits for first (second) generation leptoquark masses of $M_{\rm LQ} > 376(422)$ GeV and $M_{\rm LQ} > 319(362)$ GeV for $\beta = 1.0$ and $\beta = 0.5$ respectively.

6. Results and conclusions

ATLAS has searched for new physics in final states with jets using between 33 and 37 $\rm pb^{-1}$ of LHC data which correspond to the full data sample recorded in 2010. We are exploring a new physics regime at the TeV scale. No discrepancies with the Standard Model were found, new limits at 95% C.L. set for several models (see Table 3).

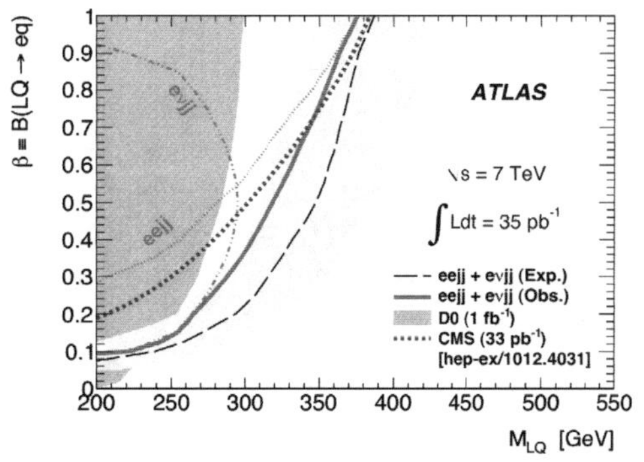

Figure 12: 95% C.L. exclusion region obtained from the combination of the two electron channels shown in the β versus M_{LQ} plane.

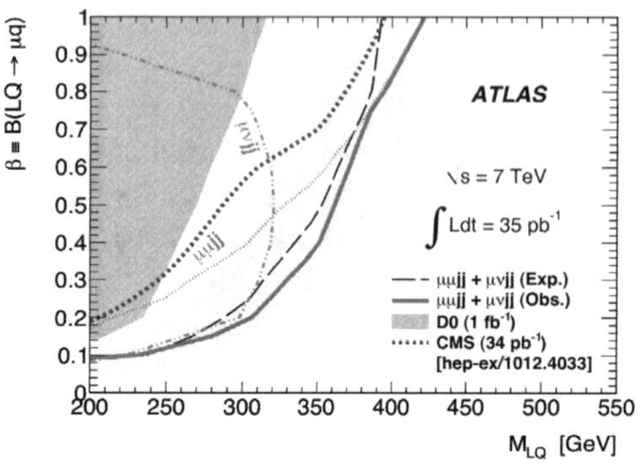

Figure 13: 95% C.L. exclusion region obtained from the combination of the two muon channels shown in the β versus M_{LQ} plane.

Table 3: Summary of exclusion limits from events with jets in the final state.

Final state	Model	Limit [TeV]
Dijets	Excited quarks ($q^\star$)	2.64
	QBH n = 6	3.67
	Axigluons	2.10
	Contact InteractionsΛ	9.50*
WWjj	4th generations heavy quarks (Q_4)	0.270
Leptons (MET) + jets	1st generation LQ ($\beta = 1$)	0.376
	2nd generation LQ ($\beta = 1$)	0.422

*6.7 TeV Bayesian Limit

References

[1] G. Aad *et al.* [ATLAS Collaboration], JINST 3 (2008) S08003.
[2] The pseudorapidity is defined in terms of the polar angle θ as $\eta = -\ln\tan(\theta/2)$.
[3] S. van der Meer, CERN-ISR-PO-68-31, 1968. URL: http://cdsweb.cern.ch/record/296752.
[4] G. Aad *et al.* [ATLAS Collaboration], ATLAS-CONF-2011-011; arXiv:1101.2185v1, accepted by EPJC. URL: http://cdsweb.cern.ch/record/1334563.
[5] G. Aad *al.* [ATLAS Collaboration], Phys. Lett. B, 694:327, 2011.
[6] G. Aad *et al.* [ATLAS Collaboration], ATLAS-CONF-2011-047, 2011. URL: http://cdsweb.cern.ch/record/1338578.
[7] G. Aad *et al.* J. Phys. **13** (2011) 053044.
[8] T. Aaltonen *et al.* [CDF Collaboration], Phys. Rev. **D79** (2009) 112002 [arXiv:0812.4036 [hep-ex]].
[9] G. Aad *et al.* [ATLAS Collaboration], Phys. Rev. Lett. **105** (2010) 161801 [arXiv:1008.2461 [hep-ex]].
[10] T. Aaltonen *et al.* [CDF Collaboration], Phys. Rev. **D79**, (2009) 011101, [arXiv:0809.3781 [hep-ex]].
[11] P. Meade and L. Randall, JHEP05 **2008**, 003 (2008). [arXiv:0708.3012 [hep-ph]].
[12] G. Aad *et al.* [ATLAS Collaboration], ATLAS-CONF-2011-022, 2011. URL: http://cdsweb.cern.ch/record/1336751.
[13] G. Aad *et al.* [ATLAS Collaboration], CERN-PH-EP-2011-050, 2011 [arXiv:1104.4481 [hep-ex]]. URL: http://cdsweb.cern.ch/record/1347253.

Supersymmetry Searches in multijet events with CMS

Raffaele Tito D'Agnolo for the CMS Collaboration

Scuola Normale Superiore, Piazza dei Cavalieri 7, 56126, Pisa, Italy
raffaele.dagnolo@sns.it

Abstract. We describe the searches for Supersymmetry performed by the CMS collaboration in events with jets and significant missing transverse energy, characteristic of the decays of heavy, pair-produced squarks and gluinos. The results presented here have been obtained with the first 36 pb^{-1} collected by the CMS detector during the 2010 data taking period.

1. Introduction

One of the many challenges, that confront us at the LHC, is the lack of a precise intuition of how physics beyond the Standard Model will manifest itself. Even if we consider only a specific class of models, such as supersymmetric theories, many possibilities are left open in terms of experimental signatures. Furthermore these signatures might depend on the fine structure of the new particles spectrum. Therefore the CMS collaboration has designed inclusive analyses based on event topology. In addition to that a great effort has been made to construct data-driven methods to estimate the Standard Model (SM) backgrounds. This has led to the development of new techniques and the production of reliable limits on new physics observables.

In these early stages of the LHC running the most promising channels for the discovery of supersymmetry involve the strong production of squarks and gluinos. If we assume R - parity conservation, their decays will be characterized by the presence of jets and high missing momentum. There are three CMS analyses designed to be sensitive to these final states. The so called α_T [1] and $\mathrm{MH_T}$ [2] analyses exploit mainly the high $\mathrm{MH_T}$ in the signal events, that we define as the magnitude of the negative vectorial sum of the p_Ts of the jets in the event. The third analysis [3], that we will call the Razor, adopts a different approach studying less traditional observables that take advantage of the kinematical properties of the pair production of the supersymmetric partners. The main SM backgrounds to these searches are: $Z(\rightarrow \nu\nu)$+jets, W+jets, $t\bar{t}$+jets and multijet events with large missing momentum from leptonic decays of heavy flavor hadrons inside jets, jet energy mismeasurement or instrumental noise and dead components.

After a brief description of the CMS detector in section 2, we examine the strategies adopted by the three all hadronic analyses to distinguish the production of supersymmetric particles from these backgrounds. In section 4 we review some of the data-driven techniques used to estimate the residual SM contribution in the signal regions and in the last section we show the results obtained with the first 36 pb^{-1} collected by the CMS detector during the 2010 running.

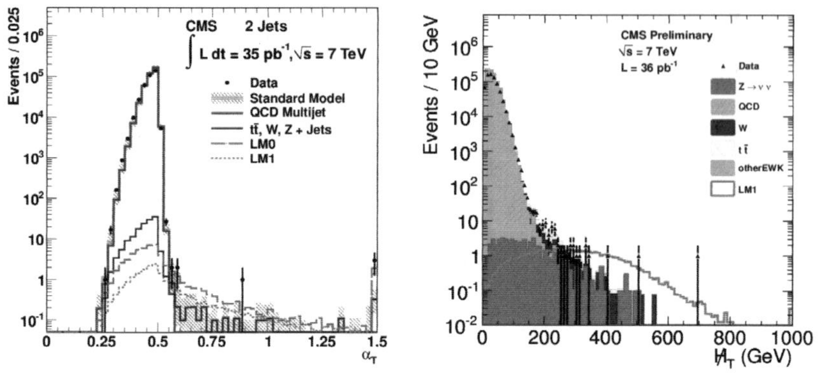

(a) (b)

Figure 1. 1(a) The α_T distribution after $H_T > 350$ GeV selection for dijet events. 1(b) $\mathrm{MH_T}$ distribution reconstructed with the Particle Flow algorithm [5].

2. The CMS detector

The Compact Muon Solenoid has been described in detail in [4]. At its core is placed a 3.8 Tesla superconducting solenoid in which silicon strips and pixels, that constitute the main tracking apparatus, are contained. The charged particles momenta can be measured with a full azimuthal coverage and up to $|\eta| < 2.5$, where η is the pseudorapidity, defined as $\eta = -\ln(\tan(\theta/2))$ and θ is the polar angle of the particle trajectory with respect to the counterclockwise proton beam direction. Outside of the solenoid the steel return yoke is instrumented with gas detectors used to identify muons.

A lead-tungstate crystal electromagnetic calorimeter (ECAL) and a brass/scintillator hadron calorimeter (HCAL) surround the tracking volume and cover the region $|\eta| <3$. An iron/quartz-fiber hadronic calorimeter in the forward region ($3 < |\eta| < 5$) contributes to the quasi-hermeticity of the detector to allow energy-balance measurements in the plane normal to the beam direction.

3. Analysis Strategies

The three CMS hadronic searches for supersymmetry do not rely on any assumption on the spectrum of the supersymmetric particles. The α_T and $\mathrm{MH_T}$ analyses explore the tails of the $\mathrm{MH_T}$ with two very different approaches towards jets mismeasurement and detector effects, while the Razor takes advantage of the properties of pair production of heavy particles. In this section we would like to describe the ideas behind the different strategies without entering into the details of each analysis, for which we refer to CMS papers and public physics analysis summaries in the bibliography.

3.1. α_T

In this search the variable α_T, inspired by [6] and first introduced in [7], was used as the main discriminator between events with real and fake missing transverse energy. Events with at least two calorimetric jets with a cone parameter of 0.5, clustered using the anti-k_T algorithm [8] and with $\mathrm{E_T} > 100$ GeV, are selected, requiring the sub-leading jets to have at least an $\mathrm{E_T}$ of 50 GeV. Additional cuts on the $\mathrm{H_T}$ and the consistency of the jet based estimate of the $\mathrm{ME_T}$ are applied and a QCD-killer cut, $\alpha_T > 0.55$, is performed. This last cut is particularly effective in reducing the background, since for perfectly measured dijet events α_T=0.5, while an imbalance of the E_T's can only make α_T smaller than 0.5. Advantages and drawbacks of this approach emerge clearly from figure 1(a), where it is easy to see how the cut on α_T is suitable for a fast discovery. On one hand it allows to ignore to a large extent the difficulties associated to cleaning the $\mathrm{ME_T}$ tails and is very effective in reducing the backgrounds, on the other it is not very efficient for typical mSUGRA [9–12] signals.

3.2. $\mathrm{MH_T}$

This analysis focuses on final states with at least three jets and high missing momentum. The baseline selection that defines this class of events consists in asking for

- ≥ 3 Jets with $|\eta| < 2.5$ and $p_T > 50$ GeV
- $\mathrm{H_T} > 300$ GeV
- $\mathrm{MH_T} > 150$ GeV
- $\Delta\phi(\mathrm{Jet_{1,2}}, \mathrm{MH_T}) > 0.5$
- $\Delta\phi(\mathrm{Jet_3}, \mathrm{MH_T}) > 0.3$
- Plus event cleaning cuts

where we have indicated with $\mathrm{Jet_{1,2}}$ the two leading jets and with $\mathrm{Jet_3}$ the additional jet required. In order to suppress the high $\mathrm{MH_T}$ tails of QCD multijets events, all the jets with $|\eta| < 5$ and $p_T > 30$ GeV are used in its measurement.

After the baseline selection two search regions are defined

(i) High $\mathrm{H_T}$, where $\mathrm{H_T} > 500$ GeV

(ii) High $\mathrm{MH_T}$, where $\mathrm{MH_T} > 250$ GeV

All the observables mentioned above, from jets to the $\mathrm{MH_T}$ in figure 1(b) have been reconstructed using the Particle Flow algorithm [5]. The jets have a cone parameter of 0.5 and have been clustered using the anti$-k_T$ algorithm. It would be worth spending several pages on the Particle Flow technique, however for our purposes it is enough to mention that the jet energy resolution improves considerably with respect to calorimetric jets for $p_T < 100$ GeV and that the $\mathrm{MH_T}$ tails are less prone to fakes coming from jet mismeasurements. This has allowed to design the analysis retaining a good efficiency for models where the masses of the new particles are low enough to be produced with sizable yield at limited integrated luminosities.

The very high efficiency on the signal is one of the main strengths of this approach, as will emerge in section 5.

3.3. The Razor

The pair production of two heavy particles, each decaying to an unseen LSP plus jets, is studied using the idea of event hemispheres following [13]. All the reconstructed objects in each hemisphere are combined into a single mega-jet. Therefore all events are forced into a dijet-like topology. In this scenario it is natural to introduce the R-frame, which is the longitudinally boosted frame that equalizes the magnitude of the two mega-jets 3-momenta. If the two heavy particles are produced approximately at rest, the R-frame is simply the center of mass frame of the collision. In this case, the magnitude of one of the two 3-momenta is proportional to an event by event estimate of the scale M_Δ

$$|\vec{p}^{\,R}_{hem_{1,2}}| = \frac{M_\Delta}{2} = \frac{M^2_{heavy} - M^2_{LSP}}{M_{heavy}}, \tag{1}$$

so it is natural to define the new variable

$$M_R \equiv 2|\vec{p}^{\,R}_{hem_{1,2}}|. \tag{2}$$

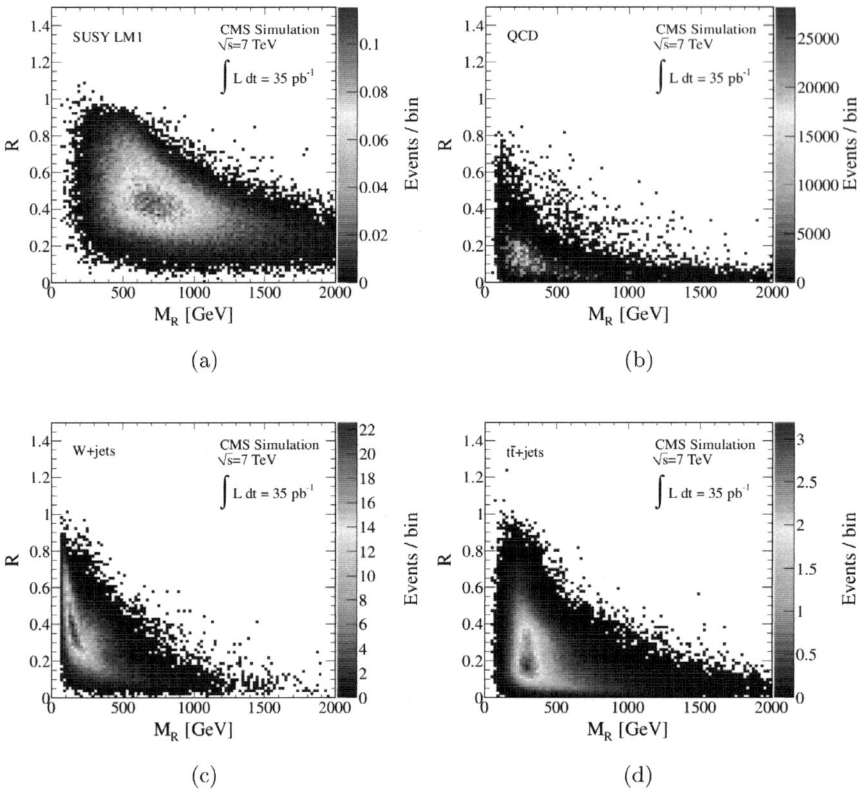

Figure 2. 2(a) Simulated SUSY events with parameter set LM1, in the (R, M_R) plane. 2(b) Simulated QCD multijet events in the (R, M_R) plane. 2(c) Simulated W+jets events in the (R, M_R) plane. 2(d) Simulated $t\bar{t}$+jets events in the (R, M_R) plane.

It is clear that M_R peaks around a characteristic scale. For example a QCD dijet event has $M_R \sim \sqrt{\hat{s}}$, whereas the production of a W boson, recoiling against a multijet system, lives in the vicinity of $M_R \sim M_W$.

The definition of M_R will indeed prove very useful, but it relies only on a subset of the full kinematical information in the event. If we use also the unseen LSPs as a source of background rejection we can define another kinematical observable as

$$R \equiv \frac{M_T^R}{M_R}, \tag{3}$$

where M_T^R is the transverse variable:

$$M_T^R \equiv \sqrt{\frac{|ME_T|(p_T^{hem_1} + p_T^{hem_2}) - M\vec{E}_T \cdot (\vec{p}_T^{\,hem_1} + \vec{p}_T^{\,hem_2})}{2}}. \tag{4}$$

Since M_T^R is a transverse version of M_R, it can be proven that R peaks at 0.5 and has its endpoint at M_Δ. In figure 2 we show the distribution of SM and SUSY events in the (R, M_R) plane. Our intuitions about R and M_R are confirmed and it is clear that a pair of cuts in that plane would be enough to obtain a very pure signal region, without further requirements on

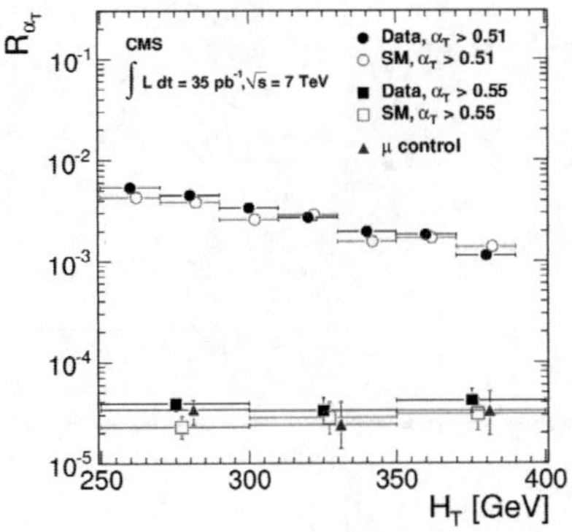

Figure 3. α_T ratio vs H_T for α_T cut values of 0.51 and 0.55.

other observables. We will review the chosen values for the cuts while presenting the results. In the next section we will see how these two variables allow a fully data-driven determination of the surviving background.

4. Background determination

All the analyses devoted to supersymmetric searches at CMS have designed independent data-driven methods to estimate the backgrounds in the signal region. These techniques will be more and more performing as the statistics increases. In this way our future sensitivity will grow much faster than $\sqrt{\int Ldt}$. In addition to that, these data-driven methods are intrinsically easy to adapt to different searches, as we will see in the following.

In the next subsection we will review the main strategies used for the three analyses, without aiming to completeness, but rather trying to give a general picture.

4.1. α_T

The total background is estimated from two control regions of low H_T. The method is based on the ratio R_{α_T} of events which pass and fail the full selection requirements. This ratio can be extrapolated from the two low H_T regions to the signal region, by virtue of its independence from H_T,

$$\frac{R_{\alpha_T}(HT300)}{R_{\alpha_T}(HT250)} = \frac{R_{\alpha_T}(HT350)}{R_{\alpha_T}(HT300)} \tag{5}$$

for events with real $\mathrm{MH_T}$. This can be seen in figure 3, where R_{α_T} is shown to be constant in H_T for the electroweak backgrounds, while it decreases slowly for QCD events, where the missing energy is mostly fake.

In addition to this procedure, the W+jets and $t\bar{t}$+jets backgrounds can be determined independently from a muon sample, while the $Z(\rightarrow \nu\nu)$+Jets background can be estimated from a γ+Jets sample. The two different methods, the inclusive one based on R_{α_T} and the exclusive ones for the electroweak backgrounds, give consistent results as will be shown in section 5.

4.2. $\mathrm{MH_T}$

A number of novel techniques have been developed to measure the background for this search in a fully data-driven way. In this limited space we will only sketch briefly the Rebalance+Smear method intended to assess the QCD multijet background. The R+S method makes a prediction by starting from a sample of events consisting of seed jets. The events are produced using an

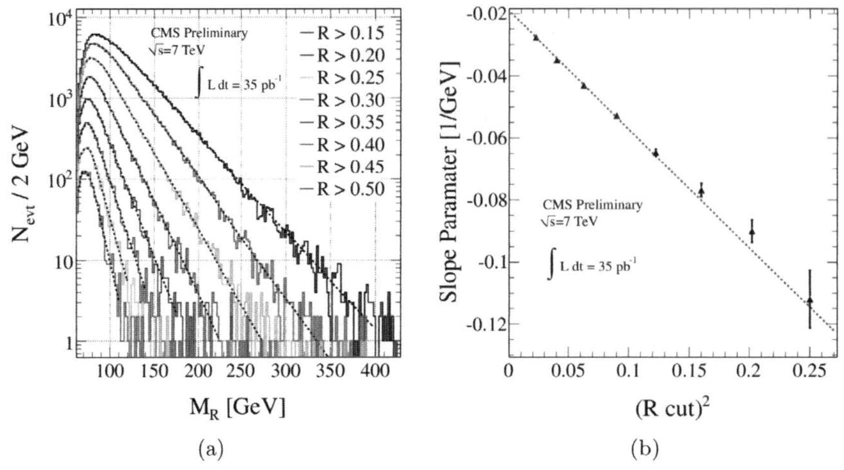

Figure 4. 4(a) M_R distribution for different values of the R threshold for events in data, selected in the QCD control box. The dotted black lines are fits to the exponential fall of the distribution. 4(b) The exponential slope S from fits to the M_R distribution, as a function of the R threshold for data events in the QCD control box.

inclusive multijet data sample as input. At first all of the jet momenta are adjusted to return the event back into approximate transverse momentum balance. Then, in the Smear step, a random value of the jet response is drawn from the jet resolution distribution for each seed jet, and its true momentum is scaled by this factor. As the input to the smearing step is representative of a pure QCD multijet sample, the resulting events can be used to apply search cuts.

As mentioned above, one of the advantages of data-driven techniques is that they are often "analysis-independent". For instance, in this case, the electroweak backgrounds have been estimated in analogy to what we have already seen for the α_T analysis. The details of the techniques are different, but the key idea of exploiting the approximate equivalence of electroweak bosons at high energies remains. So the $Z(\rightarrow)\nu\nu$+jets background can be modeled using either a γ+jets sample or a W+jets sample where the muon is treated as a neutrino.

It is not unreasonable to object to the choice of purely data-driven methods. For example in the R+S technique the jet resolution distribution must me modeled using data and nothing assures us a priori that the total error would be smaller than that coming from simply taking it from the Montecarlo. However a very thorough work has been conducted, the details of which can be found in [2] and in an upcoming paper, assuring a good sensitivity to the analysis as we will see in section 5.

4.3. The Razor

As noticed in the previous section M_R and R are strongly correlated and their correlation can be used to estimate the residual background. The M_R distribution for all the relevant backgrounds was found to fall exponentially after a turn on determined by the scale of the process under study. In general the exponential tail has a single component whose slope has a simple functional dependence on the cut on R

$$S = a + b(Rcut)^2. \qquad (6)$$

This allows to extrapolate the number of background events from regions of low M_R, where the SM is known to dominate, to the signal region. Therefore to estimate the background it is sufficient to select exclusive boxes where the single components of the SM dominate and then extrapolate from low M_R. These boxes are an hadronic one for QCD and a muon and an electron box for the electroweak backgrounds. Their complete definition can be found in [3]. In figure 4 the behaviour of M_R is shown for data, together with the results of the fit to the exponential slope.

5. Results

The three analyses have not found any significant excess with respect to the expected yield from SM processes. Therefore we can only set limits in the SUSY parameter space. As an example, we show the number of events found by the α_T analysis with the first 36 pb^{-1} collected by the CMS detector.

- In the data we are left with 13 events.
- For the backgrounds there is the inclusive prediction from R_{α_T} : $9.4^{+4.0}_{-4.8}$(stat)$\pm$1.0(syst)
- The exclusive predictions for the electroweak backgrounds. W+jets and $t\bar{t}$+jets, from a muon+jets sample: $6.1^{+2.8}_{-1.9}$(stat)$\pm$1.8(syst). $Z(\rightarrow \nu\nu)$+jets, from a γ+jets sample: $4.4^{+2.3}_{-1.6}(stat)$$\pm$1.8(syst)

We can easily see that the sum of the two electroweak components is consistent with the inclusive prediction, which is in turn compatible with the number of events from the data and with a Montecarlo prediction of 9.3$\pm$0.9(stat). The Razor and $\mathrm{MH_T}$ analyses have found the same degree of compatibility between data and SM predictions, but with different yields. The Razor with $R > 0.5$ and $M_R > 500$ GeV sees 7 events in the hadronic box and, in the muon box, for slightly relaxed cuts on the two variables, just 3.

In figure 5 we show a comparison between limits obtained by different CMS analyses. It is easy to see that the $\mathrm{MH_T}$ setup gives a limit similar to the Razor, whereas the cut on α_T affects the sensitivity of the analysis. The limit set by the $\mathrm{MH_T}$ strategy is dominated by the high $\mathrm{MH_T}$ search region defined in section 3, whereas the high $\mathrm{H_T}$ region is more prone to background contamination giving a smaller S/B.

We have shown that the CMS SUSY hadronic analyses are performing well and that their sensitivity will increase non trivially with 2011 data. In addition to that several novel background estimation methods have been designed, that might be relevant also for future searches and make us confident in the robustness of our results. In one sentence: we are ready for a supersymmetric year.

References

[1] Khachatryan V *et al.* (CMS Collaboration) 2011 *Phys.Lett.* **B698** 196–218 * Temporary entry * (*Preprint* 1101.1628)

[2] Khachatryan V *et al.* (CMS Collaboration) 2011 CMS-PAS-SUS-10-005

[3] Khachatryan V *et al.* (CMS Collaboration) 2011 CMS-PAS-SUS-10-009

[4] Adolphi R *et al.* (CMS Collaboration) 2008 *JINST* **3** S08004

[5] Khachatryan V *et al.* (CMS Collaboration) 2009 CMS-PAS-PFT-09-001

[6] Randall L and Tucker-Smith D 2008 *Phys.Rev.Lett.* **101** 221803 (*Preprint* 0806.1049)

[7] Khachatryan V *et al.* (CMS Collaboration) 2009 CMS-PAS-SUS-09-001

[8] Cacciari M, Salam G P and Soyez G 2008 *JHEP* **0804** 063 (*Preprint* 0802.1189)

[9] Chamseddine A H, Arnowitt R L and Nath P 1982 *Phys.Rev.Lett.* **49** 970

[10] Barbieri R, Ferrara S and Savoy C A 1982 *Phys.Lett.* **B119** 343

[11] Hall L J, Lykken J D and Weinberg S 1983 *Phys.Rev.* **D27** 2359–2378

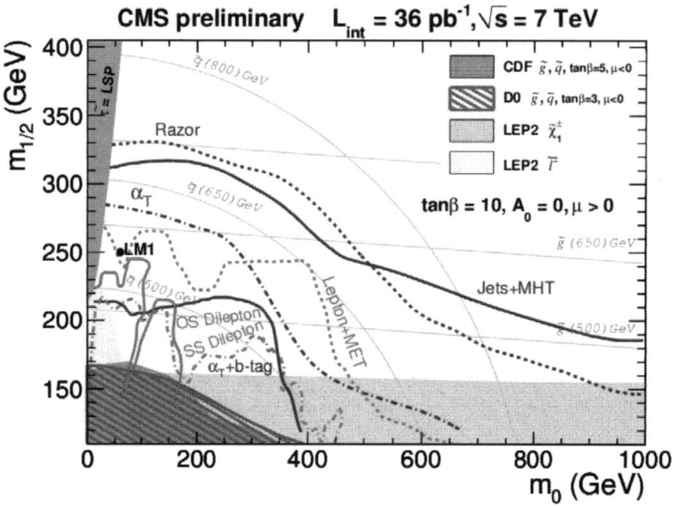

Figure 5. Observed limits from 2010 CMS supersymmetry searches in the mSUGRA $(m_0, m_{1/2})$ plane.

[12] Nath P, Arnowitt R L and Chamseddine A H 1983 *Nucl.Phys.* **B227** 121

[13] Rogan C 2010 (*Preprint* 1006.2727)

Top Physics at CDF

Gueorgui Velev
CDF Collaboration
Fermi National Accelerator Laboratory, Batavia, IL 60510, USA
E-mail: `velev@fnal.gov`

Abstract. Top quark physics is one of the most important successes of the concluding Tevatron program. In this paper, a summary of the most recent measurements of top quark properties, including the mass and its 2011 average, will be presented. Some common techniques and discussion of major systematic uncertainties for top measurements will also be presented.

1. Introduction

With the conclusion of the Tevatron Run II colliding operation in the fall of 2011, top quark physics still reminds one of the most important priorities of the CDF program. Since the first evidence in 1994 [1] and the discovery of the top quark in 1995 [2][3], CDF invested a lot of effort to determine the top quark properties, especially the top quark mass. This effort combined with the newly collected data samples gave the collaboration the ability to produce, for the first time, analyses with results which are systematically dominated, e.g. the recent top mass and $t\bar{t}$ cross-section measurements.

In the past several years, the Fermilab accelerator complex performed extremely well averaging approximately 2.25 fb^{-1} per year. Until May 2011, the Tevatron delivered more than 11 fb^{-1}. The Run II-upgraded collider detectors continued to work with a high efficiency and collected more than 9 fb^{-1}. At the end of Run II, the delivered luminosity is expected to reach 12 fb^{-1}. The current plot with integrated luminosity delivered by the Tevatron and acquired by CDF is shown at Fig. 1.

This paper reports on the latest CDF top quark results which are based on about 6 fb^{-1} acquired data until the summer of 2009. It is expected about 4 more fb^{-1}will be collected by the end of the Run II.

At the Tevatron energy of $\sqrt{s}$=1.96 TeV, top quarks are produced generally in pairs from the process $q\bar{q} \rightarrow t\bar{t}$ in ~85% of the cases. In the remaining ~15%, $t\bar{t}$ pairs are produced via gluon fusion ($gg \rightarrow t\bar{t}$). Additionally, top can be produced as a single quark by electroweak interactions, by W-gluon fusion or virtual W* production in the s-channel [4], but with a smaller cross section. This paper will not discuss the results from the single top analyses [5].

In the Standard Model the branching ratio of the decay $t \rightarrow bW$ is almost 100%. When a $t\bar{t}$ pair is produced, each of the W-bosons can decay into either a charged lepton and a neutrino (BR = 1/3) or into a $q\bar{q}^{'}$ quark pair (BR = 2/3). This allows us to classify the final states as:

- Dilepton final state (DL), when both W bosons from the $t\bar{t}$ pair decay leptonically. This

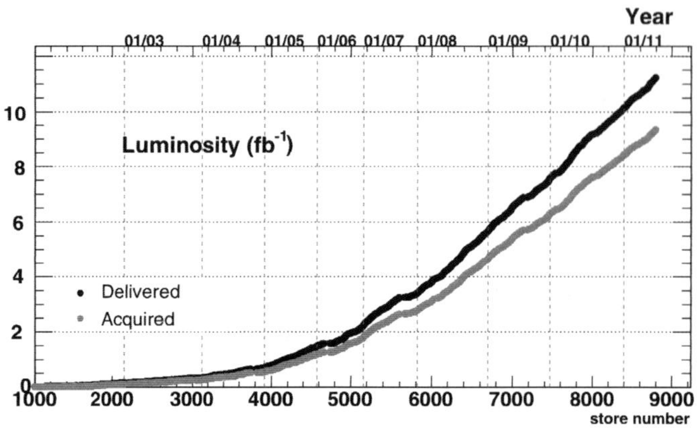

Figure 1. The Tevatron integrated luminosity and the CDF acquired luminosity until May 2011.

state is characterized by two energetic (high P_T) charged leptons, two jets from $b\bar{b}$ quarks [1] and significant missing transverse energy ($\not\!\!E_T$) from the neutrinos.

- Lepton plus jets final state ($L + J$), when one W boson decays leptonically and the other one hadronically. This state contains one high P_T charged lepton, four jets and significant $\not\!\!E_T$.
- All-hadronic final state (ALH), when both W bosons decay hadronically. This state is characterized by 6 jets, two of which are from b-quarks.
- $\not\!\!E_T$ +Jets final state ($\not\!\!E_T + j$), this is a sample of events comming mostly from $L + J$ when the high enrgy lepton is not detected due to limited detector coverage.

In Run II, CDF uses all of these signatures to measure the top quark properties, including the top quark mass.

2. Measurements of the top quark mass

As it was pointed in many previous papers and conference talks, top quark mass is a fundamental parameter of the Standard Model (SM). In the SM, the value of its mass cannot be predicted from the global electroweak fit with accuracy better than ~ 8 GeV/c^2. Thus, since the 1995 discovery, both the CDF and DØ experiments have been improving the top mass measurement accuracy.

The current understating of the origin of the non-zero masses in the SM is explained by spontaneous breaking of the electroweak symmetry induced by the Higgs field [6]. The top quark, which is the heaviest known fundamental particle in the SM with a mass value comparable to the electroweak scale, magnifies the suspect that this quark may play a special role in the mechanism of electroweak symmetry breaking. In addition, because of its mass, the top quark gives the largest contribution to loop corrections in the W propagator. Within the SM, the correlation between the top mass and the W mass induced by these corrections gives us the ability to set limits on the mass of the Higgs boson favoring a relatively light Higgs [7]. A more accurate measurement of the top quark mass will tighten the SM predicted region for the Higgs mass. Moreover, a precise measurement can be used as an additional constraint on the supersymmetric models beyond the SM [8].

[1] Errors in jet reconstruction and gluon radiation in the event may increase or decrease the observed number of jets. This statement is valid for all final states.

2.1. *Top kinematics and mass reconstruction methods*

The $t\bar{t}$ kinematics and reconstruction methods depends on the final states. Historically, it was shown that the lepton plus jets final state provides the best mass measurement sensitivity. This channel combines signatures with high signal-to-background ratio (S/B), especially in the silicon vertices tagged events, with over-constrained kinematics. This means, that the number of measured quantities in the $L+J$ channel and the number of applicable energy-conservation equations, from five production and decay vertexes, are larger than the number of non-measured kinematical parameters in an event. This feature allows for a complete reconstruction of the four-momenta of the final state particles in the event, for example using the two constraint kinematical fit ($2\ CF$) - a reconstruction of the top mass event-by-event.

In this type of analysis there is an ambiguity in how to assign the four leading jets to the two b and two light quarks coming from the $t\bar{t}$ system, so-called combinatorial uncertainty. If none of these jets is tagged by b taggers[2], there are 12 different ways of assigning jets to the 4 partons. Combining with the ambiguity from solving of a quadratic equation on P_z^{ν} of the neutrino, there are 24 different values of M_{top} returned by the kinematical fit. The combinations are reduced to 12 or to 4 if one or two of the jets are selected from b-taggers.

In contrast to the lepton plus jets mode, the di-lepton case, due to the presence of two neutrinos, has unconstrained kinematics. For this channel, the number of non-measured kinematical variables is larger by one than the number of kinematic constraints ($-1\ CF$). Obviously, it is impossible to find only one solution for the mass value per event. Typical approaches to solve the unconstrained problem is to integrate over several of the event kinematic variables. Another way to approach the problem is to assume some kinamatical event parameters ($\vec{C}$) as known in order to constrain the kinematics and then vary $\vec{C}$ to determine a set of solutions. In this case, to determine a preferred mass, one may average all of the mass solutions or select the the most probable one.

The minimal requirement in the case of $-1\ CF$ kinematics is to use a two dimensional vector as $\vec{C}$. In DL analyses, we chose the pseudo-rapidities or the azimuthal angles of the two neutrino momenta ($\vec{C} = (\eta_{\nu 1}, \eta_{\nu 2})$ or $\vec{C} = (\phi_{\nu 1}, \phi_{\nu 2})$) to create a net of solutions in the $\vec{C}$ plane.

In case of the all-hadronic final state, a $3\ CF$ kinematic fit[3] can be applied. This channel is kinematically easy to reconstruct but suffer from a very low S/B ratio, due to the large QCD multi-jet background which dominates the $t\bar{t}$ signal by 3 orders of magnitude after the application of the online trigger selection only. To improve S/B, requirements based on the kinematical and topological characteristics of standard model $t\bar{t}$ events are expressed in terms of a neural network and applied to the data. The neural network selection is followed by the requirement of jets to be b-tagged from the silicon vertex detector. This allows us to achieve a S/B of $\sim$ 1:2.

Table 1 summarizes the S/B in case of different finals states and after applying the b-jet tagging identification. Additional improvement of the S/B could be achieved if soft lepton b-jet tagger is applied [11] (it is not represented in the table).

The variety of the CDF analyses in different final states is relatively large. However, all analyses can be separated in three major categories:

(i) Template type analyses, where the reconstructed data distributions are compared with expected distributions from Monte Carlo generated signal (mass-dependent) and background. In these types of analyses, all events are weighted equally. By doing so one neglects the additional information coming from a different mass resolution in single events.

[2] CDF uses taggers based on either displaced vertices (Secondary Vertex Tagging, SVX, for example see [10]) or on low P_T electrons or muons from the b-quark semileptonic decays (Soft Lepton Tagging, SLT [11])

[3] If the conservation of the transverse momentum in the $p\bar{p}$ interaction point is applied, $5\ CF$ is possible to be performed

Table 1. Signal-to-background ratio for different final states.

Sample	Di-lepton S/B (e,μ)	Lepton plus Jets S/B (e,μ)	All-hadronic S/B after NN kin. selection	$\not{E}_T$ plus Jets S/B after NN kin. selection
0 b-tags	1:1	1:3	1:20	1;10
1 b-tags	4:1	3:1	1:3	3:2
2 b-tags	20:1	10:1	1:1	6:1
Top Events in 1 fb^{-1} ($\geq$ 1 b-tags)	25	180	300	150

(ii) Matrix Element type analyses, originally proposed by Dalitz and Goldstein [12] and independently by Kondo [13]. These methods calculate the posterior probability, given the known production cross section, for every event with measured kinematic properties, to originate from a $p\bar{p} \to t\bar{t}$ process.

(iii) A mixture between methods 1) and 2). For example the $t\bar{t}$ event is reconstructed using the kinematic algorithms similar to the template analyses but an event-by-event probability of each kinematic reconstruction is exploited as a weight (e.g. $exp(-\frac{\chi^2}{2})$). A typical example of this category is the CDF all-hadronic ideogram analysis [14].

2.2. Jet Energy Scale

In Run I, the uncertainty coming from the calorimeter jet energy scale was dominating the total systematic error. This uncertainty, commonly named *a priori* uncertainty, was determined from different calibration samples like photon plus jet, Z plus jet and di-jet events [15]. There are many sources of uncertainties related to jet energy scale at CDF, but the major ones can be summarized as:

- Relative response of the calorimeters as a function of pseudorapidity.
- Single particle response linearity in the calorimeters.
- Jet Fragmentation.
- Modeling of the underlying event energy.
- Correction for the energy deposited out of the jet cone.

The upper uncertainties are evaluated separately as a function of the jet p_T and η. The summary contributions are shown in Fig.2 for the η region of $0.2 < \eta < 06$. The black lines show the total $\pm$ 1σ total uncertainty which is taken as a unit of jet energy scale miscalibration in the analyses.

In Run II, CDF takes advantage of the larger $t\bar{t}$ samples available and employ new analysis techniques to reduce the JES uncertainty. In particular, the Run II analyses in the lepton plus jets, all hadronic and $\not{E}_T$ +Jets channels constrain the response of light-quark jets using the kinematic information from $W \to qq'$ decays (*in situ* calibration). Residual JES uncertainties associated with p_T and η dependencies as well as uncertainties specific to the response of b-jets are treated separately. This method decreased in the order of magnitude the systematic uncertainty on the top quark mass associated with JES. Moreover, in analyses where W decays only leptonically, a JES scale determined from external data samples can be used to decrease top mass uncertainty.

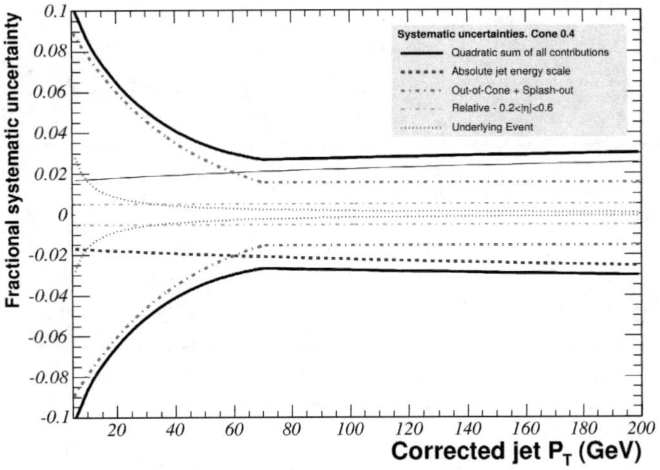

Figure 2. Jet energy scale (JES) uncertainty as a function of the corrected jet p_T for the underlying event (dotted red), relative response (dashed green), out-of-cone energy (dashed red) and absolute response (dashed blue). The contribution of all sources are added in quadrature (solid black line) and shows the total JES systematic uncertainty.

2.3. Lepton plus jets final state

The most accurate top quark mass measurement comes from the lepton plus jets final state. As it was mentioned earlier, this final state combines the benefits of good S/B ratio, the possibility to reconstruct the top quark mass event by event with a relatively small combinatorial effect, and a large branching fraction. In brief, we summarize below the main selection criteria for this channel. The events have the signature $p\bar{p} \rightarrow t\bar{t}X \rightarrow \ell\nu bq\bar{q}'\bar{b}X$. The characteristics of this final state begins with the identification of one isolated central high energy lepton (e with $E_T >$ 20 GeV or μ with $P_T >$ 20 GeV) and $|\eta| < 1$[4]. In recent analyses, CDF included the samples with forward muons $|\eta| > 1$ which are collected with a dedicated trigger on missing transverse energy [17] with a muon p_T >20 GeV/c. Assuming that the lepton is coming from W boson decay, a companion neutrino should exist. As the neutrino energy is not detected, we require $\not{E}_T$ >20 GeV in the event.

In order to fully reconstruct the $t\bar{t}$ system, at least four central jets $|\eta| \leq 2$ are required in the system. The SVX tagging algorithm is run over the leading jets (E_T >15 GeV). To obtain maximum statistical benefit from the event sample it is helpful to decompose it into two classes of events (1-SVX tag and 2-SVX tag) which are expected to have different S/B ratios and therefore different top mass resolutions.

The recent most accurate CDF $L + J$ analysis is based only on the sample with four tight jets [18] and integrated luminosity of 5.6 fb^{-1}, and uses the Matrix element methodology.

The dominant backgrounds in all samples are direct W plus multi-jet production, including heavy flavor production, and QCD multi-jet events where one jet is misidentified as a lepton. Additional small backgrounds are due to WW/WZ and single top production, Table 2.

For each event, a likelihood function depending on of the true top mass M$_{top}$ and deviation from the true detector energy scale ΔJES is constructed as:

$$L_{ev}(\vec{y} \mid \mathrm{M}_{top}, \Delta\mathrm{JES}) = \frac{1}{N}\sum_{1}^{24}\int \frac{PDF(z_p, z_{\bar{p}})}{Flux}\ \mathrm{TF}(\vec{y} \mid \vec{x}, \Delta\mathrm{JES})\ |\mathrm{ME}(\mathrm{M}_{top}, \vec{x})|^2\ \Phi(\vec{x}) \qquad (2.1)$$

[4] A complete description of the lepton selection, including all cuts used, can be found elsewhere [16]

Table 2. Expected sample composition for an integrated luminosity of 5.6 fb^{-1}. The $t\bar{t}$ contribution is estimated using a cross section of 7.4 pb [19] and M_{top} = 172.5 GeV/c^2.

Event	1 b-tag	2 b-tag	Source/Generator
$W + Heavyflavor$	129.5±42.2	15.7±5.5	Alpgen[20]
Non-W (QCD)	50.1±25.5	5.5±3.8	Data[21]
$W + Lightflavor$ (Mistag)	48.5±17.1	1.0±0.4	Alpgen
Diboson (WW,WZ,ZZ)	10.5±1.1	1.0±0.1	Pythia[22]
Single top	13.5±0.9	4.0±0.4	Madgraph[23]
$Z \to ll$ + jets	9.9±1.2	0.8±0.1	Pythia
Total background	261.8±60.6	28.0±9.6	-
$t\bar{t}$ signal	767.3±97.2	276.5±42.0	Pythia
Total expected	1029.1±114.5	304.5±44.1	-
Total observed	1016	247	Data

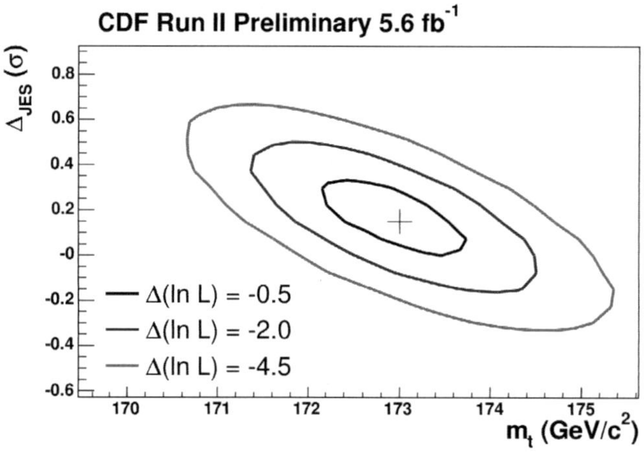

Figure 3. 2D likelihood on the 1087 data events: the contours corresponding to 1σ, 2σ, and 3σ uncertainty in the M_{top} measurement are shown.

where $\vec{y}$ are the kinematical quantities measured in the detector (e.g. jets and lepton momenta), $\vec{x}$ is the parton level variables which define the event kinematics, N is a normalization factor including the event acceptance, PDF are the parton distribution functions for incoming $p\bar{p}$ with momentum fractions z_p and $z_{\bar{p}}$, $Flux$ is the relativistic flux factor, and $\mathrm{TF}(\vec{y} \mid \vec{x}, \Delta\mathrm{JES})$ are the transfer functions that estimates the probability specific parton kinematic to produce measured jets. $\mathrm{ME}(M_{top}, \vec{x})$ is the leading order matrix element and it includes both $q\bar{q} \to t\bar{t}$ and $gg \to t\bar{t}$ production processes with appropriate spin correlations. All 24 possible permutations are summed with weight corresponding to the probability jet to originate from b or light quark.

The total likelihood L_{total} for all candidate events are calculated under the assumption that all events are signal. In reality, L_{total} contains contributions from both signal and background events. To correct for the background events, an average background contribution to L_{total} is calculated and subtracted.

Fig. 3 shows the L_{total} 2D likelihood contours for 1σ, 2σ and 3σ. The next step is to treat ΔJES as a nuisance parameter and eliminate it using the profile likelihood method, taking the maximum value of L_{total} along the ΔJES axis for each M_{top}. This procedure gives us ability to extract the top mass value of 173.0 ± 0.9 GeV/c^2. Two effects are contribution to the error of 0.9 GeV/c^2. The first part is the statistical uncertainty on M_{top} and the second part is the uncertainty due to ΔJES. By fixing the ΔJES to correspond to the maximum value of L_{total}, one can separate these contributions: 173.0 ± 0.7 $(stat.) + 0.6$ (JES) GeV/c^2.

Table 3 shows typical systematic categories for top mass analyses. These categories and how they are evaluated are the outcome of many years of effort of the CDF and DØ collaborations [24]. Many of the systematic uncertainties decreased significantly with time due to increasing knowledge of the source of the data versus MC discrepancies, adding improved particle generators, and new calibration samples. For CDF mass analyses, even after applying of the *in situ* calibration, the residual JES uncertainty associated with the p_T and η dependencies still dominates in the systematic tables. This fact stresses the importance of better calorimeter calibrations in wide p_T and η regions.

The other systematic uncertainties in Table 3 correspond to sources from: the calibration of the method, signal Monte Carlo modeling (differences between PYTHIA and HERWIG [25] generators), tuning the parameters used for initial state radiation and final state radiation to CDF Drell-Yan data, additional uncertainties on JES for b-jets (reflecting to the differences between light and gluon jets versus b-jets), uncertainty on the lepton p_T scale, multiple hadron interactions (we take into account uncertainty on the jet corrections as a function of the number of interactions in the event), uncertainties arising from the PDFs, the background modeling and the systematic uncertainty due to color reconnection effects. The individual systematic uncertainties added in quadrature is 0.88 GeV/c^2. This determines the total top quark mass uncertainty and makes the analysis the most precise single measurement available to date: 173.0 ± 0.9 $(stat. +$ JES$) \pm 0.9$ $(syst.)$ GeV/c^2

2.4. Dilepton Final State

CDF has a dilepton final state measurement of the top quark using 5.6 fb^{-1} of data. This measurement was performed utilizing the template methodology. The analysis uses two

Table 3. Typical systematic uncertainties on M_{top}. The categories are the same for all top quark mass analyses. The values are from ME $L + J$ analysis.

Systematic source	Systematic uncertainty (GeV/c^2)
Residual JES	0.49
Calibration	0.10
MC generator	0.37
ISR and FSR	0.15
b-JES	0.26
Lepton P_T	0.14
Multiple hadron interactions	0.10
PDFs	0.14
Background modeling	0.34
Gluon fraction	0.03
Color reconnection	0.37
Total	0.88

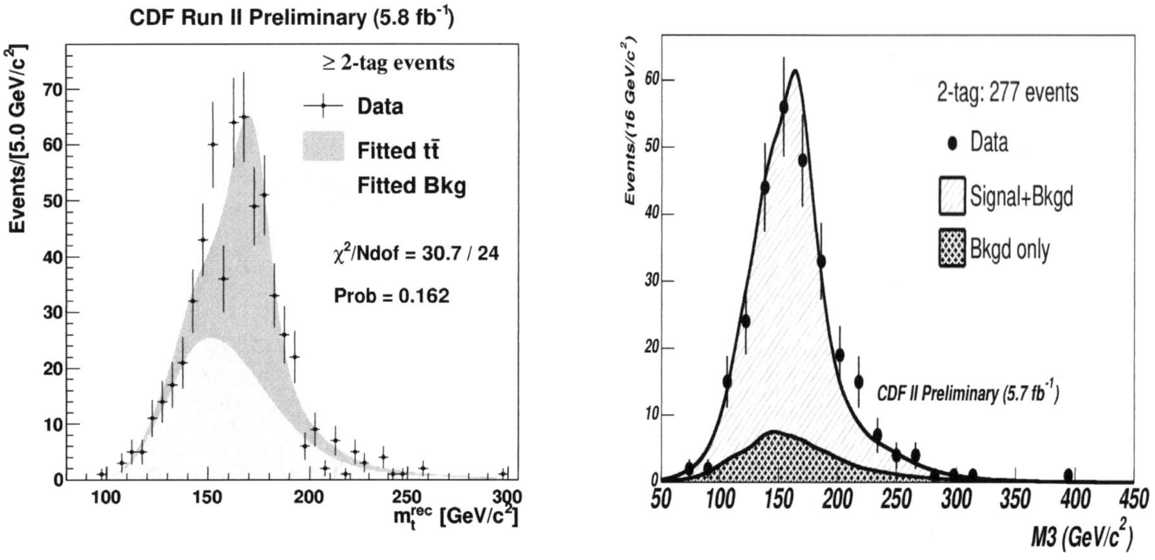

Figure 4. Reconstructed M_{top} in the case of *ALH* for events with two b-tag jets is shown on the left. The right plot shows the reconstructed invariant mass *M*3, described in the text, for the events with two b-tag jets in the $\not{E}_T$ +Jets final state.

dimensional templates where one of the two variables is M_{top} reconstructed with the neutrino weighting method [26, 27]. The idea of the method is to scan over the unknown pseudorapidities of the two neutrinos and to weight the solutions based on the $\not{E}_T$ event resolution. The second variable is m_{T2} [28, 29] which corresponds the the transverse mass in two missing particles final states, in our case the two neutrinos. A total of 237 events with no tagged jets and 155 with at least one tagged jet are selected with expected background of 82±22 and 9±1 events respectively. One can see that non-tagged and tagged events have very different S/B and using them as statistically independent samples increases the sensitivity of the method. This analysis provides the current most accurate published measurement in *DL* of 170.3±2.0 (*stat.*)±3.1 (*syst.*) GeV/c^2 [30].

2.5. *All-hadronic final state*

As it was discussed in 2.1, *ALH* has the advantages of a large BR, and of a fully reconstructable kinematics - all partons from the $t\bar{t}$ could be detected in the detector. The major downside is the huge background from QCD multi-jet production.

The most recent and precise measurement of M_{top} in this channel has been obtained by the CDF experiment using template methodology with 5.8 fb^{-1} of data [31]. After pre-selection cuts and a requirement of at least one b-tagged jet, a neural net is applied. This neural net includes both kinematical and jet shape variables increasing the S/B ratio to 1:3 for single b-tagged events and 1:1 for the double and more b-tagged events. Then, a kinematic fit is performed to reconstruct, for each event, the top and W masses. These variables are then used as templates to fit the data in order to obtain the M_{top} and JES, simultaneously.

A total of 2256 events with exactly one b-tagged jet and 600 with at least two b-tagged jets are selected, with an expected background of 1712±77 and 305±22 events respectively. The background is estimated directly from the non-tagged events, where the correction for the small amount of signal is applied. The final likelihood fit returns 172.5 ± 1.7 (*stat.* + JES) ± 1.2 (*syst.*) GeV/c^2 corresponding to an accuracy of 1.2%. The data fit in case of two b-tagged events is shown in Fig. 4, left.

2.6. $\not{E}_T$ +Jets final state
For the first time CDF performed analysis in a final state called $\not{E}_T$ +Jets [32] with 5.7 fb^{-1} data. This final state contains events mostly from the lepton plus jets decay mode where the lepton is not reconstructed due to limited detector coverage. To start the selection we use fully hadronic trigger with high jet multiplicities, at least four jets. Additionally, the events are required to have a well measured $\not{E}_T$ but no leptons. The last requirement is an important criteria in creating a statistically independent sample. The next step is to apply a neural network selection to improve S/B in this sample. The events with lower probability to be produced by the $t\bar{t}$ interactions are used to estimate the shape of the QCD background.

The fact that this analysis does not identify the lepton from $t\bar{t}$ event makes it impossible to perform a full kinematical reconstruction. However, we can reconstruct the invariant mass called $M3$, which is defined as the invariant mass of three jets that give the largest jet $\not{E}_T$. To increase the chance that these three jets come from the same top quark decay, we select two of the three jets without b-tags to have an invariant mass m_{jj} closest to the W mass and use them to constrain JES in the analysis. We also reconstruct a third variable called $M3'$, similar to $M3$ mentioned above, except that the third jet is different from any of the three jets that construct $M3$. A multidimensional likelihood fit with three dimensional templates ($M3$, $M3'$ and m_{jj}) is performed. The final result from the likelihood is 172.3 ± 2.4 ($stat.+$JES) ± 1.0 ($syst.$) GeV/c^2 corresponding to an accuracy of 1.5% in the determination of the top mass. Fig. 4, right, shows the reconstructed $M3$ variable for double b-tagged events at the top mass of 172.5 GeV/c^2.

2.7. Top mass combination
CDF combined the measurements from different channels using the best linear unbiased estimation technique [33]. For CDF measurements, the method returned the most probable mass $M_{top} = 172.7 \pm 1.1$ GeV/c^2 [34]. By combining with DØ results [35], we obtained the best Tevatron top quark mass of 173.2 ± 0.9 GeV/c^2. For first time the total uncertainty is below 1.0 GeV/c^2 and the relative precision is 0.5%. Fig. 5, left, shows the Run I and Run II measurements included in the latest CDF combination. All of these measurements are statistically independent; the systematic uncertainties are combined according to the correlations. The plot on the right extrapolates how the top mass uncertainty is expected to decrease with increasing of the integrated luminosity starting from the first CDF statistically significant measurement at 680 pb^{-1}. The triangles represent the measurements or averages of M_{top} up to 5.8 fb^{-1}. The solid curve represents the trend which keeps all the systematic uncertainties at level of the 680 pb^{-1} measurement, but scales the statistical and "in-situ" JES uncertainties with the integrated luminosity. The dotted curve scales the total uncertainty with luminosity. It represents the lower bound on how well the top mass could be measured. CDF measurements are close to the dotted curve showing a large improvement of the systematic uncertainty with the integrated luminosity.

3. Measurements of the top quark properties
Additionally to the determination of the top mass, we have studied the top quark properties in various ways using its unique characteristics. Top quark decays before hadronization thus the property information is carried by the its decay products. In this paper we present a selection set of measured top properties. The full set of measurement could be found in the CDF public web-pages [36].

A recent and intriguing CDF result with 5.3 fb^{-1} of data is the measurement of the forward-backward asymmetry A_{FB} in the $t\bar{t}$ production. This measurements can be done only at the Tevatron and cannot be reproduced at LHC where the initial state is symmetric. In LO QCD, the top production angle is symmetric with respect to the beam direction. In NLO QCD, a small charge asymmetry, $A_{FB} = 0.05 \pm 0.01$, is predicted and it is due to the to

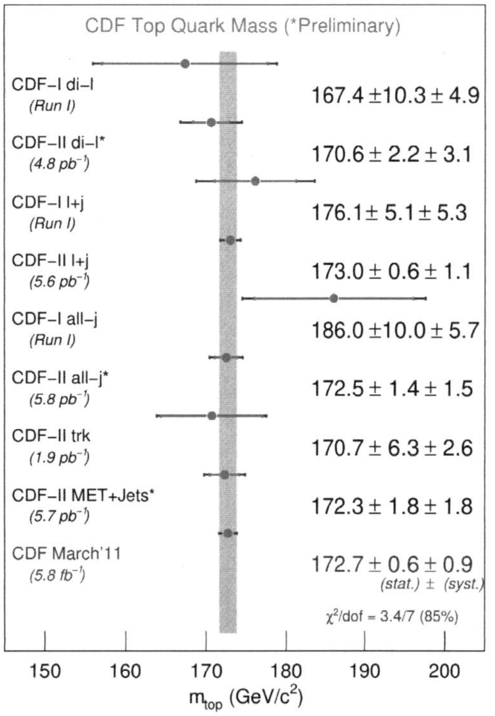

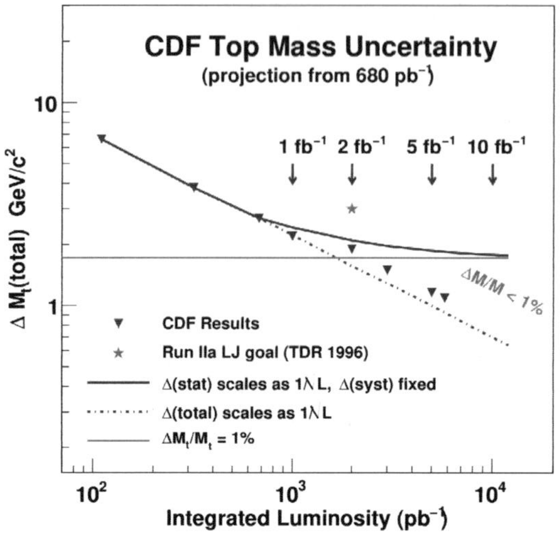

Figure 5. The combination of the top mass is shown in the left plot. The right plot shows the expected uncertainty corridor versus the integrated luminosity and the CDF measurements.

interference between initial-state radiation diagrams with final-state ones [37]. The integral forward-backward asymmetry A_{FB} is defined as $A_{FB} = \frac{N_{top}(p)-N_{top}(\bar{p})}{N_{top}(p)+N_{top}(\bar{p})}$ where $N_{top}(p)$ $(N_{top}(\bar{p}))$ is the number of tops observed in the proton (anti-proton) direction. CDF measured A_{FB} in $L+J$ and DL channels. Both results deviate 2σ, $A_{FB}^{LJ} = 0.181 \pm 0.075(stat. + syst.)$ and $A_{FB}^{DL} = 0.42 \pm 0.16(stat. + syst.)$, compared to the SM model prediction. Moreover, for the $L+J$ analysis CDF investigated the dependence of A_{FB} on the rapidity y and the $t\bar{t}$ invariant mass $M_{t\bar{t}}$. If some discrepancy in the dependence of A_{FB} is observed a possible interpretation could be explained with the presence of extra particles decaying to $t\bar{t}$ with a large forward-backward asymmetry. Fig. 6 shows the measured asymmetry dependence on $M_{t\bar{t}}$. While in the low region (< 450 GeV$/c^2$) measurement A_{FB} is consistent with the SM prediction, a 3σ deviation is observed in the high mass region. The rapidity y dependence of A_{FB} shows a 2σ deviation from the SM prediction.

Top quark is so heavy that it is practically impossible directly to determine its lifetime. However, one can measure the top decay width which is proportional to the top lifetime. CDF has a direct measurement of the top quark width Γ_t based on 4.3 fb^{-1} $p\bar{p}$ collision. This analysis uses template methodology and exploits the dependence of the reconstructed M$_{top}$ width in the $L+J$ channel on Γ_t. Each event is compared to templates created with different Γ_t and the most probable value is extracted. The correct accuracy do not allow us to measure Γ_t but rather to find a limit. By applying a Feldman-Cousins approach, we establish an upper limit of $\Gamma_t < 7.6$ GeV at 95% CL and a two-sided 68% CL interval of 0.3 GeV $< \Gamma_t <$ 4.4 GeV [38].

The $t\bar{t}$ spin correlation is well predicted by the SM and any deviation is a sign for new physics coupled to the top quark. The spin state can be observed in the angular correlations in the DL channel where the two leptons are well measured. In this channel, CDF found $k = -0.086^{+0.59}_{-0.57}$ which is consistent with standard model prediction of $k \sim 0.78$ within 95% CL interval [39]. This measurement is currently limited by the statistics of the data sample.

The SM predicts that the W→tb vertex is a 100% V-A coupling. As a consequence, $\sim$

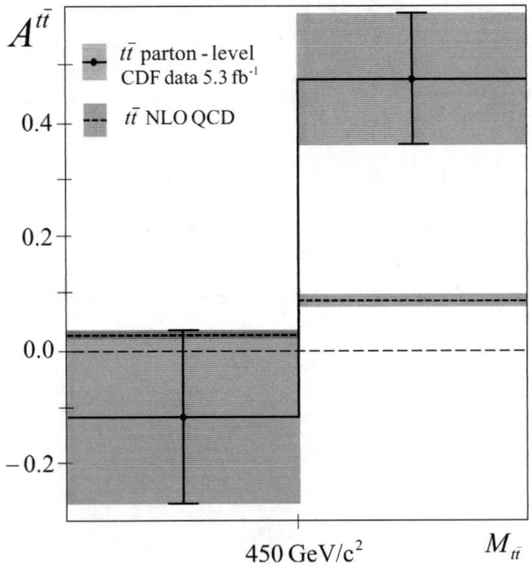

Figure 6. Measured forward-backward asymmetry A_{FB} for two regions of the $t\bar{t}$ invariant mass $M_{t\bar{t}}$

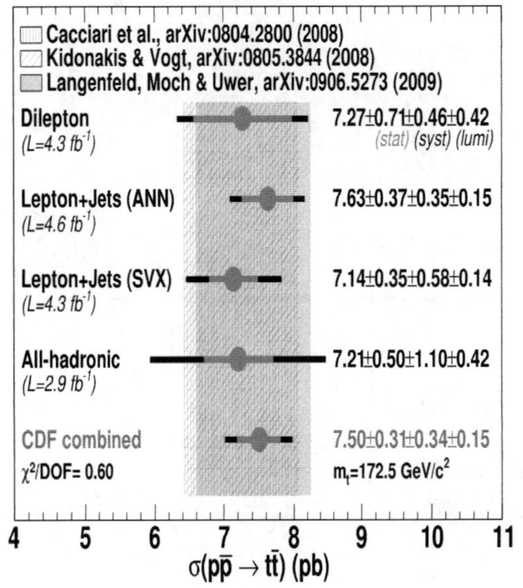

Figure 7. Summary of the CDF $t\bar{t}$ cross section measurements and the combined value.

70% of W bosons are longitudinally (f_0) polarized and in the other 30% of the cases they have a left handed polarization (f_-). Any new particles, which may be involved in the same decay topologies and having non-standard couplings could create a different ratio of polarized W bosons, even producing a few right-handed (f_+) ones. CDF has measurements for the W polarization in the $L + J$ [40] and DL [41] final states. In the most recent one, a model independent simultaneous measurement of f_0 and f_+ yields $f_0 = 0.78 \pm 0.20(stat.) \pm 0.06(syst.)$ and $f_+ = -0.12 \pm 0.11(stat.) \pm 0.04(syst.)$ which are also in good agreement with the SM.

Flavor changing neutral current (FCNC) are predicted in SUSY and other exotic physics models. In the SM, top FVNC decay modes are highly suppressed; in CDF any measurable signal from FCNC decays will indicate an evidence of new physics. FCNC decay of top quark (e.g. $t \to Zq$) predict different final states comparing to the $t\bar{t}$ event where $t \to Wb$. For example, if $Z \to l^+l^-$ and $W \to q\bar{q}'$ one could expect a dilepton final state without large $\not{E}_T$. CDF performed this type of analysis and set the 95% CL upper limit at 3.7% [42].

Finally, CDF measured $t\bar{t}$ cross section in all possible channels. The summary of these measurements and combined cross section value of $\sigma = 7.50{\pm}0.31(stat.){\pm}0.34(syst.){\pm}0.15(lum.)$ are shown in Fig. 7.

4. Conclusion

We presented the current status of the CDF top quark analyses. Due to the excellent performance of the Tevatron collider and the CDF detector, and many improvements in the analysis techniques, the collaboration has been performed many challenging measurements revealing the top quark properties [36]. Some of the measurements started to be systematically dominated. For first time, the Tevatron combined top quark mass has a precision less than 1 GeV/c^2 or $\frac{\Delta \mathrm{M}_{top}}{\mathrm{M}_{top}}$ =0.5% and the $t\bar{t}$ cross section with a precision of 6.5%. By the end of Run II, September 2011, we expect to have 12 fb^{-1} delivered to the CDF experiment, which approximately doubles our data analysis set. New and improved measurements are expected and the possibility of surprising results is not excluded.

References

[1] Abe F *et al.* 1994 Phys. Rev. D **50** 2966 and Phys. Rev. Lett. **73** 225
[2] Abe F *et al.* 1995 Phys. Rev. Lett. **74** 2626
[3] Abachi S *et al.* 1995 Phys. Rev. Lett **74** 2632
[4] Willenbrock S D, Dicus D A 1986 Phys. Rev. D **34** 155
[5] Aaltonen T *et al.* 2009 Phys. Rev. D **82** 112005
[6] Higgs P W 1964 Phys. Lett. **12** 132
[7] The Tevatron Electroweak Working Group 2010 FERMILAB-TM-2466-E
[8] Heinemeyer S, Kraml S, Porod W, Weiglein G 2003 *Physics Impact of a Precise Determination of the Top Quark Mass at an* e^+e^- *Linear Collider*, arXiv:hep-ph/0306181
[9] Quigg C 1997 FERMILAB-Pub-97/091-T
[10] Acosta D *et al.* 2005 Phys. Rev. D **71** 052003.
[11] Aaltonen T *et al.* 2009 Phys. Rev. D **79** 052007
[12] Dlaitz R H, Goldstein G R 1999 Proc. R. Soc. London A **445** 2803
[13] Kondo K 1998 J. Phys. Soc. **57** 4126
[14] Aaltonen T *et al.* 2007 Phys. Rev. Lett. **98** 142001
[15] Bhatti A *et al.* 20026 Nucl. Instrum. Methods Phys. Res., Sect. A **566** 375
[16] Acosta D *et al.* 2005 Phys. Rev. Lett. **94** 091803
[17] Acosta D *et al.* 2005 Phys. Rev. D **72** 052003
[18] Aaltonen T *et al.* 2010 Phys. Rev. Lett. **105** 252001
[19] Moch S, Uwer P 2008 Nucl. Phys. B, Proc. Suppl. **183** 75
[20] Mangano M *et al.* 2003 J. High Energy Phys. **07** 001
[21] Acosta D *et al.* 2005 Phys. Rev. D **71** 072005
[22] Sjöstrand T *et al.* 2001 Comput. Phys. Commun. **135** 238
[23] Maltoni F, Stelzer T 2003 J. High Energy Phys. **02** 027
[24] Aaltonen T *et al.* 2009 Phys. Rev. D **79** 072001
[25] Corcella G *et al.* 2001 J. High Energy Phys. **01** 010
[26] Abbott B *et al.* 1999 Phys. Rev. D **60** 052001
[27] Abulencia A *et al.* 2006 Phys. Rev. D **73** 112006
[28] Lester C, Summers D 1999 Phys. Lett. B **463** 99
[29] Barr A, Lester C, Stephens P 2003 J. Phys. G **29** 2343
[30] Aaltonen T *et al.* 2011 Phys. Rev. D **83** 111101
[31] Aaltonen T *et al.* 2011 CDF Collaboration Public Note **10456**
[32] Aaltonen T *et al.* 2011 CDF Collaboration Public Note **10433**
[33] Lyons L, Gibaut D, Clifford P 1988 Nucl. Instrum. Meth. **A270** 110
[34] Aaltonen T *et al.* 2011 CDF Collaboration Public Note **10444**
[35] Aaltonen T *et al.* 2011 CDF Collaboration Public Note **10549**
[36] *More information about CDF measurements of top quark properties can be found at http://www-cdf.fnal.gov/physics/new/top/top.html*
[37] Almeida L G, Sterman G, Vogelsang W 2008 Phys. Rev. D **78** 014008
[38] Aaltonen T *et al.* 2010 Phys. Rev. Lett. **105** 232003
[39] Aaltonen T *et al.* 2009 CDF Collaboration Public Note **9824**
[40] Aaltonen T *et al.* 2010 Phys. Rev. Lett. **105** 042002
[41] Aaltonen T *et al.* 2010 CDF Collaboration Public Note **10333**
[42] Aaltonen T *et al.* 2008 Phys. Rev. Lett. **101** 192002

Top mass reconstruction in ATLAS

Andrea Neusiedl[1] on behalf of the ATLAS Collaboration
[1] University of Mainz, Staudingerweg 7, 55128 Mainz, Germany
E-mail: `neusiedl@uni-mainz.de`

Abstract. The top-quark mass is a fundamental parameter of the Standard Model. After the discovery of the top quark, the measurements of its properties were of substantial interest. Within the framework of the SM, the top-quark mass can be used in combination with other electroweak precision measurements to constrain the mass of the yet unobserved Higgs boson. In the new era of the Large Hadron Collider (LHC), the first top quarks have been produced in Europe in proton-proton collisions at a center-of-mass energy of 7 TeV. The top-quark mass measurement of ATLAS in the so called lepton+jets channel with 35 pb^{-1} integrated luminosity is presented. In this early data-taking period the largest uncertainty on this measurement comes from the knowledge of the jet energy scale. It is shown how this uncertainty is determined and which methods are used for measuring the top-quark mass.

1. Introduction

After the discovery of the top-quark in 1995, much work has been devoted to precisely measure its mass. The present world average value is $m_{\text{top}} = (173.3 \pm 1.1)$ GeV [1] [1]. Until now the most precise measurements are based on the $t\bar{t} \to$ lepton+jets channel with lepton $= e, \mu$, where one of the W bosons decays leptonically and the other hadronically. The largest systematic uncertainty results from the limited accuracy of the jet energy scale (JES).
Especially in first ATLAS data this will be the dominating contribution to the total systematic uncertainty. In ATLAS three complementary methods in the $t\bar{t} \to$ lepton+jets channel are exploited and a brief summary is given here. A detailed description can be found in [2].

- a 2-dimensional template analysis (2d-analysis) that simultaneously determines m_{top} and a global Jet energy Scale Factor (JSF).
- a 1-dimensional template analysis (1d-kinfit analysis) exploiting a kinematical likelihood fit to all decay products.
- a 1-dimensional template analysis (1d-R_{32} analysis) which is based on the measurement of the mass ratio:

$$R_{32} \equiv \frac{m_{\text{top}}^{\text{reco}}}{m_{\text{W}}^{\text{reco}}}. \tag{1}$$

  $m_{\text{top}}^{\text{reco}}$ and $m_{\text{W}}^{\text{reco}}$ are the per event reconstructed invariant masses of the top-quark and the W boson, respectively.

With the presently available luminosity and the current understanding of the ATLAS data the 1d-R_{32} analysis yields the smallest expected uncertainty and is chosen as the present baseline and presented in the following two sections.

[1] Natural units are used throughout, i.e. $c = \hbar = 1$.

2. Data sample, object definitions and event selection

The data used were collected with the ATLAS detector [3] at the LHC. The integrated luminosity of the data sample amounts to $\mathcal{L}_{\text{int}} = 35.3$ pb^{-1} with an uncertainty of 3.4% [4].
The sources of background considered are: $t\bar{t}$ events in all other decay channels, single top-quark production processes, di-boson processes, W/Z+jets processes, and finally QCD multijet production, which is the only source of background which is directly determined from data [5]. The $t\bar{t}$ signal samples were generated with the MC@NLO program [6, 7]. The single top-quark production events are simulated with the MC@NLO code, the di-boson processes with the HERWIG program [8], and the W/Z+jets events with the ALPGEN [9] generator.
The SM top-quark decays into a W boson and a b-quark in almost 100% of the cases. The event is expected to contain a charged lepton, missing transverse energy ($E_{\text{T}}^{\text{miss}}$) from the undetected neutrino and four jets with high transverse momentum, two of which are b-jets. Both lepton channels are kept separate throughout the analysis and the $t\bar{t} \rightarrow$ electron+jets ($t\bar{t} \rightarrow$ muon+jets) channel is referred to as electron (muon) channel. Events are selected by applying the following requirements to the reconstructed objects:

- **Trigger:** It is required that an electron or muon must have fired the ATLAS trigger. The triggers used are fully efficient for leptons with $p_{\text{T}} > 20$ GeV.
- **Primary vertex:** A primary vertex with at least five charged particles assigned to it is required.
- **Electrons:** Electron candidates are defined by the following criteria: $|\eta| < 1.37$ or $1.52 < |\eta| < 2.47$, $p_{\text{T}} > 20$ GeV, and the electron must be isolated, see [5]. No high-p_{T} and isolated muon must be present[2].
- **Muons:** Muon candidates are defined by the following criteria: $|\eta| < 2.5$, $p_{\text{T}} > 20$ GeV, and as in the case of electrons, muons must be isolated, and the separation to the closest jet with $p_{\text{T}} > 20$ GeV has to be larger than $\Delta\text{R} = 0.4$. No high-p_{T} and isolated electron must be present.
- **Missing transverse energy $\mathbf{E}_{\text{T}}^{\text{miss}}$:** The $E_{\text{T}}^{\text{miss}}$ is calculated from the vector sum of the transverse energy of muons and all calorimeter cells. A requirement of $E_{\text{T}}^{\text{miss}} > 35$ GeV (20 GeV) is used in the electron (muon) channels to help separate events with and without leptonic W boson decays.
- **W boson transverse mass $\mathbf{m}_{\text{W}}^{\text{T}}$:** In the electron channel $m_{\text{W}}^{\text{T}} > 25$ GeV is required, where m_{W}^{T} is calculated from the momentum vectors of the charged lepton candidate and the missing transverse energy.[3] In the muon channel a triangular region in the $E_{\text{T}}^{\text{miss}} - m_{\text{W}}^{\text{T}}$ plane is excluded by requiring $E_{\text{T}}^{\text{miss}} + m_{\text{W}}^{\text{T}} > 60$ GeV.
- **Jets:** Jet candidates are reconstructed with the anti-k_{t} algorithm [10, 11] with a distance parameter of 0.4. The jets are calibrated to the hadronic energy scale. Jet candidates must have $|\eta| < 2.5$ and $p_{\text{T}} > 25$ GeV. Events must contain at least four jets. At least one jet is required to be b-tagged by the SV0 [5] algorithm. Events containing a jet with $p_{\text{T}} > 20$ GeV which fails basic jet quality criteria are rejected. Jets are removed from the event if they are closer than $\Delta\text{R} = 0.2$ to any selected electron.

In addition to the selection listed above, two requirements specific to the 1d-R_{32} analysis are imposed. If the jet-triplet assigned to the hadronic top-quark contains two b-tagged jets the event is rejected. Furthermore, the reconstructed mass of the W boson jet candidates must be

[2] In the ATLAS right-handed coordinate system the x-axis points towards the center of the LHC ring, the y-axis points upwards and the z-axis points in the direction of the counter-clockwise running proton beam. The polar angle θ and the azimuthal angle ϕ are defined with respect to the z-axis and x-axis, respectively. The pseudo rapidity is defined as $\eta = -\ln(\tan(\theta/2))$ and the radial distance in (η, ϕ) space is $\Delta\text{R} = \sqrt{\Delta\eta^2 + \Delta\phi^2}$.

[3] Transverse mass is given by: $M_{\text{T}} \equiv (E_{\text{T}}(\text{lepton}) + E_{\text{T}}^{\text{miss}})^2 - (p_{\text{T}}(\text{lepton}) + E_{\text{T}}^{\text{miss}})^2$

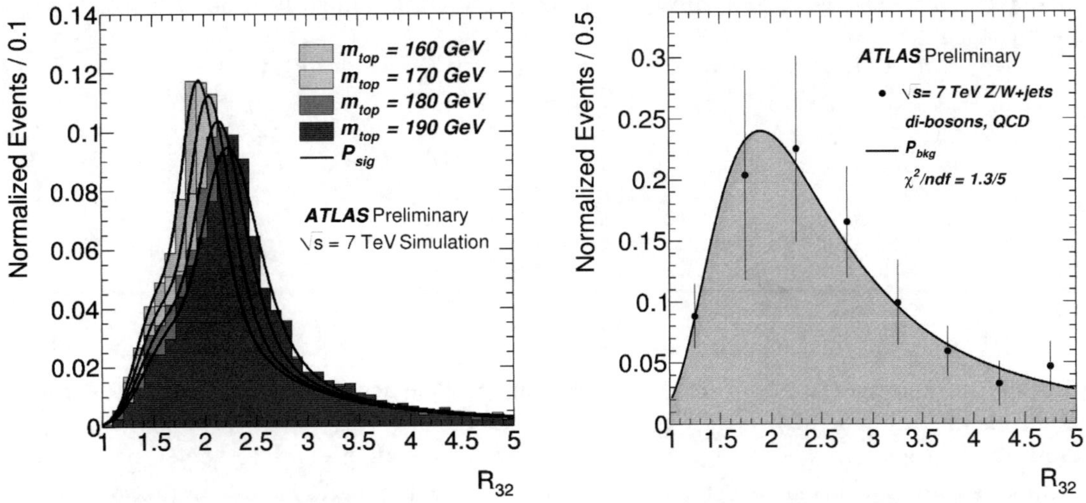

(a) Signal, electron channel

(b) Background, electron channel

Figure 1. The signal templates (a), together with their respective fits for four $m_{\rm top}$ values are shown (only the electron channel). The background template (b) is shown with their respective fits (only the electron channel).

within a window of 60 GeV $< m_{\rm W}^{\rm reco} <$ 100 GeV. Consequently, the numbers of events used by the complementary analyses are correspondingly larger than in the baseline method. With this selection the expected background contribution is 15% (16%) for the electron (muon) channel, and the corresponding signal-to-background ratio (S/B) is 5.8 (5.2).

3. Results from the 1d-R_{32} analysis

The R_{32} method uses the ratio of the reconstructed 2-jet and 3-jet invariant masses and is therefore inherently more stable against JES variations.

3.1. Main concept

In the following the main elements of this method are described: the template parametrizations and the likelihood construction. For each of the events selected the hadronic top-quark mass is reconstructed. Among all selected jets, the three whose 4-vector sum yields the highest transverse momentum are identified as the decay products of the hadronically decaying top-quark. If this jet-triplet contains exactly one b-tagged jet the other two jets are assigned to the W boson. If the jet-triplet contains no b-tagged jets, the association of jets to the W boson is performed by choosing the pair of jets with the smallest ΔR in the reconstructed top-quark candidate rest-frame. The signal templates are constructed from Monte Carlo samples for different $m_{\rm top}$ assumptions [160, 170, 172.5, 180, 190] GeV. The signal templates are described by the sum of a ratio of two Gaussians and a Landau function, the latter represents mainly wrong jet-triplet assignments. In Figure 1(a) the sensitivity to $m_{\rm top}$ is visualized by showing the superposition of the signal templates and their fits for four $m_{\rm top}$ values in the electron channel. For the background the $m_{\rm top}$ independent parts are treated together and their R_{32} distribution is described by a Landau function. In Figure 1(b) the background template and its fit is shown for the electron channel. The functions choosen describe the templates adequately and also for the muon channel the templates are equally well described.

These parametrizations (PDFs) are used in an extended unbinned likelihood fit to the data:

$$\begin{aligned}
\mathcal{L}(R_{32}|m_{\text{top}}) &= \mathcal{L}_{\text{shape}}(R_{32}|m_{\text{top}}) \times \mathcal{L}_{\text{n}_\text{s}+\text{n}_\text{b}} \times \mathcal{L}_{\text{bkg}} , \qquad (2)\\
\mathcal{L}_{\text{shape}}(R_{32}|m_{\text{top}}) &= \prod_{i=1}^{N} \frac{n_{\text{s}} \cdot P_{\text{sig}}(R_{32}|m_{\text{top}})_i + n_{\text{b}} \cdot P_{\text{bkg}}(R_{32})_i}{n_{\text{s}} + n_{\text{b}}} ,\\
\mathcal{L}_{\text{n}_\text{s}+\text{n}_\text{b}} &= \frac{e^{-(n_\text{s}+n_\text{b})} \cdot (n_\text{s}+n_\text{b})^N}{N!} ,\\
\mathcal{L}_{\text{bkg}} &= \exp\left\{ -\frac{(n_\text{b} - n_\text{b}^{\text{pred}})^2}{2\sigma_{n_\text{b}^{\text{pred}}}} \right\} .
\end{aligned}$$

The likelihood function (eq. 2) contains three terms. $\mathcal{L}_{\text{shape}}$ accounts for the shape of the R_{32} distribution and its dependence on the top-quark mass m_{top}, $\mathcal{L}_{\text{n}_\text{s}+\text{n}_\text{b}}$ determines the relative contribution of signal (n_s) and background (n_b) events to all observed events, N, and $\mathcal{L}_{\text{bkg}}$ constrains the normalization of background events within the background prediction of n_b^{pred} events. The validity of this approach is verified with the pseudo-experiment technique. A good linearity is found between the input top-quark mass ($m_{\text{top}}^{\text{in}}$) used to perform the pseudo-experiments, and the result of the fits, averaged over 50 pseudo-experiments.

3.2. Evaluation of systematic uncertainties

To evaluate systematic uncertainties pseudo-experiments are performed and the systematic difference is evaluated with respect to the standard analysis. Neglecting possible correlations, the total uncertainty is calculated as the quadratic sum of all individual contributions.

The sources of systematic uncertainties considered are the following:

Method calibration: The limited statistics of the Monte Carlo samples leading to a systematic uncertainty for the template fits.

Signal Monte Carlo generator: The systematic uncertainty related to the choice of the generator which influences the signal template shapes, compared are MC@NLO and POWHEG samples [12].

Hadronization: Choice of hadronization model (Lund-String [13] or the Cluster model [8]).

Pileup: Systematic uncertainty from the imperfect description of pile-up in MC compared to data.

Initial (ISR) and final state (FSR) QCD radiation: Systematic uncertainty from different amounts of ISR and FSR altering the observed jet multiplicity. Dedicated ACERMC [14, 15] signal samples with modified QCD parameter Λ are used.

Proton parton density function: The signal samples are generated using the CTEQ66 [16] proton parton density function. The impact of the uncertainties of the proton parton density functions on R_{32} are evaluated.

W/Z+jets background normalization: The influence of the limited knowledge of the normalization is taken into account by altering the Monte Carlo predicted background fraction by $\pm 100\%$.

W/Z+jets background shape: The impact of the variation of the shape of the W/Z+jets background contribution based on changes observed on stable particle jets while re-weighting Monte Carlo model parameters.

QCD background normalization: The QCD normalization is varied by $\pm 100\%$.

QCD background shape: As suggested in [5] two shape variations per decay channel are applied and the fit is repeated.

Jet energy scale: Impact on R_{32} of jet energy scale variations (average JES uncertainty about $\pm 4\%$).

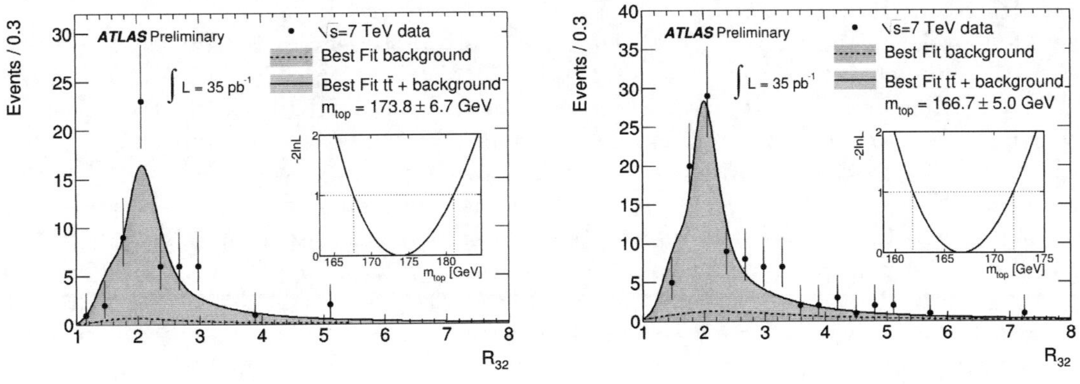

(a) Electron channel (b) Muon channel

Figure 2. The measured R_{32} distribution for the electron (a) and muon (b) channels. In addition the background PDF and the total PDF are shown for the measured top-quark mass, i.e. for the top-quark mass corresponding to the minimum of the likelihood function as shown in the inset (only statistical uncertainty given).

Relative b-jet to light jet energy scale: Impact on R_{32} of a varied b-jet energy scale of 2.5%.
b-tagging efficiency and mistag rate: The b-tagging efficiency and mistag rates in data and Monte Carlo simulation are not identical. The uncertainty on those b-tagging scale factors which describe these differences are taken into account.
Jet energy resolution: The uncertainty of the jet energy resolution is evaluated by smearing each reconstructed jet in simulation and re-doing the event selection and fitting.
Jet reconstruction efficiency: The jet reconstruction efficiency in data and Monte Carlo are found to agree within an accuracy of 2%. To account for this difference, jets are randomly removed from the events.
The three biggest contributions to the systematic uncertainty are the ISR/FSR uncertainties (about 2.5 GeV), the b-jet energy scale (2.5 GeV) and the jet energy scale (about 2 GeV). The W/Z+jets background normalization contributes of the order of 1.5 GeV. The other uncertainties are of the order or below 1 GeV. The numbers given represent a mean value of both channels. The total systematic uncertainty is 4.8 GeV in the electron channel and 5.0 GeV in the muon channel.

3.3. Top-quark mass measurement

Figure 2 shows the results of the R_{32} analysis when performed on data. For both channels the fit function adequately coincides with the distribution observed in the data. The results from both channels are statistically consistent with each other, and they are also in agreement with the present world average for m_{top}. The values measured of the top-quark mass are:

$$\begin{aligned} m_{\text{top}} &= (173.8 \pm 6.7 \pm 4.8)\ \text{GeV} \quad \text{electron channel} \\ m_{\text{top}} &= (166.7 \pm 5.0 \pm 5.0)\ \text{GeV} \quad \text{muon channel} \\ m_{\text{top}} &= (169.3 \pm 4.0 \pm 4.9)\ \text{GeV} \quad \text{combined} \end{aligned}$$

4. Results from the 2d-analysis

In the 2d-analysis, similarly to [17], both m_{top} and a global Jet energy Scale Factor (JSF) are determined simultaneously by using the $m_{\text{top}}^{\text{reco}}$ and $m_{\text{W}}^{\text{reco}}$ distributions. This analysis performs an in-situ jet rescaling and the systematic uncertainty on m_{top} stemming from the JES is reduced and partly transformed into an additional statistical uncertainty on m_{top} due to the 2-d fit, see [18]. For this a global JSF (averaged over η and p_{T}) is obtained, which is mainly based on the observed differences between the predicted $m_{\text{W}}^{\text{reco}}$ distribution and the data. The well known values of m_W and Γ_W are used to improve on the experimental resolution of $m_{\text{top}}^{\text{reco}}$. For the events fulfilling the common requirements listed in Section 2 a kinematic fit using the hadronic W boson decay candidates is performed. For each event the two jets of each light jet pair with a reconstructed $m_{\text{W}}^{\text{reco}}$ within 30 GeV ($\approx 3\sigma$) around the peak position of the $m_{\text{W}}^{\text{reco}}$ distribution are input to the following χ^2:

$$\chi^2 = \left[\frac{E_{j1}(1-\alpha_1)}{\sigma_1}\right]^2 + \left[\frac{E_{j2}(1-\alpha_2)}{\sigma_2}\right]^2 + \left[\frac{M_{jj}(\alpha_1,\alpha_2) - m_W}{\Gamma_W}\right]^2 \quad (3)$$

This fit determines the best light jet combination in the event and the corresponding parton scale factors (α_{i}) for their jet energies. The jet-triplet assigned to the hadronic top-quark decay is then obtained by adding the b-tagged jet which maximizes the p_{T} of the jet-triplet. From these jets the two observables $m_{\text{W}}^{\text{reco}}$ and $m_{\text{top}}^{\text{reco}}$ are constructed. The $m_{\text{W}}^{\text{reco}}$ is calculated using the reconstructed light jet 4-vectors (i.e. not applying the α_{i}) of the combination with the smallest χ^2. In contrast, $m_{\text{top}}^{\text{reco}}$ is calculated from these light jet 4-vectors scaled to the parton level (i.e. applying the α_{i}) and the above determined b-tagged jet. After the χ^2 fit, the observed numbers of events are 134 (210) in the electron (muon) channel. The respective predicted numbers of signal events are 112.4 ± 0.7 and 153.7 ± 0.8, and the S/B amounts to 4.0 and 4.3, in the electron and muon channel. Templates are constructed for $m_{\text{top}}^{\text{reco}}$ as functions of an input top-quark mass in the range $[160-190]$ GeV, and input values for the JSF in the range $[0.9-1.1]$, and, finally, for $m_{\text{W}}^{\text{reco}}$ as a function of the assumed JSF for the same range. Pseudo-experiments also here verified the unbiased reconstruction of $m_{\text{top}}^{\text{reco}}$.

For the muon channel the observed $m_{\text{top}}^{\text{reco}}$ distribution and the corresponding fit are shown in Figure 3.

The results for the two channels and their combination are:

$$\begin{aligned} m_{\text{top}} &= (168.3 \pm 6.2 \pm 4.3)\ \text{GeV} \quad \text{electron channel} \\ m_{\text{top}} &= (163.5 \pm 6.7 \pm 4.6)\ \text{GeV} \quad \text{muon channel} \\ m_{\text{top}} &= (166.1 \pm 4.6 \pm 4.4)\ \text{GeV} \quad \text{combined} \end{aligned}$$

The measured JSF is $1.08\,^{+0.04}_{-0.06}$ ($1.01\,^{+0.05}_{-0.05}$) in the electron (muon) channel, uncertainties quoted are statistical. This analysis has a reduced JES systematic of 0.7 GeV but a reduced statistical precision. The other important uncertainties, i.e. those from the relative b-jet to light jet energy scale and ISR/FSR variations, are about the same size as for the baseline analysis.

5. Results from the 1d-kinfit analysis

This analysis explores a kinematical likelihood fitter to relate the observed objects to parton level predictions. The entire reconstructed objects are input to the kinematical fitter. In this procedure the measured jets relate to the quark decay products of the W boson and the top-quarks, and the $E_{\text{T}}^{\text{miss}}$ is identified with the transverse momentum components of the neutrino. The kinematical likelihood is defined as:

$$\text{L} = \text{BW}(\hat{\text{E}}_{\text{jet},1}, \hat{\text{E}}_{\text{jet},2}|\text{m}_{\text{W}}, \Gamma_{\text{W}}) \cdot \text{BW}(\hat{\text{E}}_{\ell}, \hat{\text{E}}_{\nu}|\text{m}_{\text{W}}, \Gamma_{\text{W}}) \cdot$$

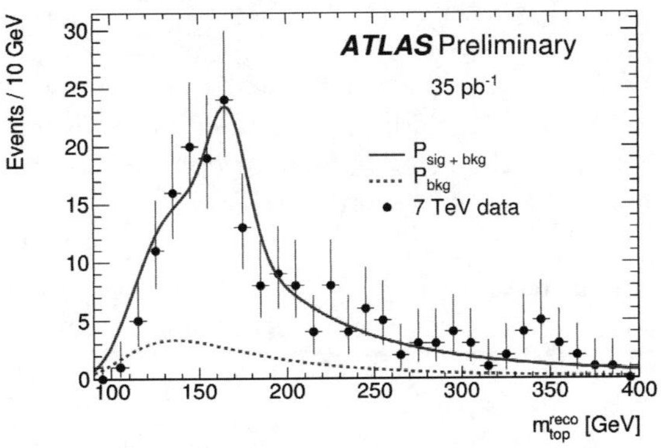

Figure 3. 2d-analysis: The reconstructed top-quark mass in data together with the fitted background PDF and the full PDF (in the muon channel).

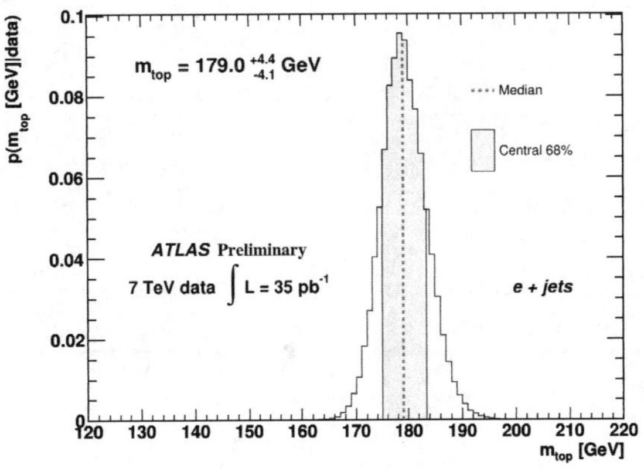

Figure 4. 1d-kinfit analysis: The likelihood profile as a function of the top-quark mass (in the electron channel).

$$\begin{aligned}
&\mathrm{BW}(\hat{\mathrm{E}}_{\mathrm{jet},1},\hat{\mathrm{E}}_{\mathrm{jet},2},\hat{\mathrm{E}}_{\mathrm{jet},3}|\mathrm{m}_{\mathrm{top}}^{\mathrm{reco}},\Gamma_{\mathrm{top}})\cdot\mathrm{BW}(\hat{\mathrm{E}}_{\ell},\hat{\mathrm{E}}_{\nu},\hat{\mathrm{E}}_{\mathrm{jet},4}|\mathrm{m}_{\mathrm{top}}^{\mathrm{reco}},\Gamma_{\mathrm{top}})\cdot \\
&\mathrm{TF}(\mathrm{E}_{\mathrm{x}}^{\mathrm{miss}}|\hat{\mathrm{p}}_{\mathrm{x},\nu})\cdot\mathrm{TF}(\mathrm{E}_{\mathrm{y}}^{\mathrm{miss}}|\hat{\mathrm{p}}_{\mathrm{y},\nu})\cdot\mathrm{TF}(\mathrm{E}_{\ell}|\hat{\mathrm{E}}_{\ell})\cdot \\
&\prod_{\mathrm{i}=1}^{4}\mathrm{TF}(\mathrm{E}_{\mathrm{jet,i}}|\hat{\mathrm{E}}_{\mathrm{jet,i}})\cdot\prod_{\mathrm{i}=1}^{4}\mathrm{TF}(\eta_{\mathrm{jet,i}}|\hat{\eta}_{\mathrm{jet,i}})\cdot\prod_{\mathrm{i}=1}^{4}\mathrm{TF}(\phi_{\mathrm{jet,i}}|\hat{\phi}_{\mathrm{jet,i}})\cdot\delta(b\text{-tagged jet}\mid b\text{-quark}).
\end{aligned} \tag{4}$$

The parton level objects are marked with a hat like $\hat{E}_{\mathrm{jet},1}$, m_W is the known W boson mass, and $m_{\mathrm{top}}^{\mathrm{reco}}$ is the estimator for the top-quark mass, i.e. the per event result. For a given event a likelihood is calculated for each possible assignment. The permutation selected maximizes the negative log likelihood. The transfer functions (TF) are derived from the signal Monte Carlo. These functions are obtained for energies and angles of light jets and b-jets, the energy of the charged lepton, and the two components of the $E_{\mathrm{T}}^{\mathrm{miss}}$. In addition, the kinematical likelihood exploits the known values of m_W and Γ_W to constrain the reconstructed leptonic and hadronic W boson masses and constrains the reconstructed leptonic and hadronic top-quark masses to be identical. The inclusion of the b-tagging information as a delta distribution restricts the number of permutations investigated per event. With this analysis the correlation of all objects is utilized leading to a narrower $m_{\mathrm{top}}^{\mathrm{reco}}$ distribution than what is achieved by the baseline analysis, resulting in a significantly reduced expected statistical uncertainty of m_{top}. However, the analysis is subject to a stronger JES sensitivity than the other two analyses described above.

The analysis is performed for the events fulfilling the common requirements listed in Section 2. The observed numbers of events after the kinematical likelihood fit are 157 (247) in the electron (muon) channel. The respective predicted numbers of signal events are 131.3±0.9 and 190.4±1.0, and the S/B amounts to 3.8 and 4.3, in the electron and muon channel. Again, pseudo-experiments verified that the mean values and widths of the pull distributions are consistent with the expectations of zero and one. The result of a likelihood fit as a function of the top-quark mass is reported in Figure 4. The maximum of the likelihood profile determines the measured m_{top}, and the 68% area around that value gives the statistical uncertainty. The results for the two channels and their combination, performed as for the baseline analysis above, are:

$$\begin{aligned}
m_{\mathrm{top}} &= (179.0 \pm 4.3 \pm 7.5)\ \mathrm{GeV} \quad \text{electron channel} \\
m_{\mathrm{top}} &= (172.0 \pm 3.5 \pm 7.5)\ \mathrm{GeV} \quad \text{muon channel}
\end{aligned}$$

$$m_{\text{top}} = (174.8 \pm 2.7 \pm 7.5) \text{ GeV} \quad \text{combined}$$

As expected the statistical precision is significantly improved in comparison to the presented baseline method. The dominating uncertainty stems from the JES and amounts to about 6.6 GeV. The other important uncertainties, i.e. those from the relative b-jet to light jet energy scale and ISR/FSR variations are about the same size as for the baseline analysis.

6. Comparison of the three top-quark mass measurements

The results for all analyses, and for each lepton channel are consistent within uncertainties. For all analyses, the muon channel has a slightly smaller expected statistical uncertainty. The systematic uncertainty is similar for both channels. The expected statistical uncertainty is smallest for the 1d-kinfit analysis, followed by the 1d-R_{32} analysis, with the 2d-analysis having the least precise statistical component. At the present level of understanding of the JES, the order of precision of systematic uncertainty is reverted. The fact that the complementary methods lead to consistent results gives confidence in their stability.

7. Summary

The top-quark mass has been measured directly via three complementary implementations of the template method in the $t\bar{t} \rightarrow$ lepton+jets channel based on first pp-collision data at $\sqrt{s} = 7$ TeV from ATLAS. The integrated luminosity used is about $\mathcal{L}_{\text{int}} = 35$ pb^{-1}. All analyses lead to consistent results within their uncertainties. The measured ATLAS value presented here is obtained from a one dimensional template analysis exploiting a new variable R_{32}, defined as the event-by-event ratio of $m_{\text{top}}^{\text{reco}}$ and $m_{\text{W}}^{\text{reco}}$. The main sources of uncertainty stem from the jet energy scale uncertainty for light jets and b-jets, and from the modelling of initial and final state radiation. Combining both channels the top-quark mass is measured with a total uncertainty of 6.3 GeV to be

$$m_{\text{top}} = (169.3 \pm 4.0 \pm 4.9) \text{ GeV},$$

where the first uncertainty is statistical, and the second one systematic.

References

[1] The CDF and D0 Collaborations and the Tevatron Electroweak Working Group 2010 (*Preprint* arXiv:1007.3178)
[2] The ATLAS Collaboration 2011 *ATLAS-CONF-2011-033*
[3] The ATLAS Collaboration, G Aad *et al.* 2008 *J. Inst.* **3** S08003
[4] The ATLAS Collaboration 2011 *ATLAS-CONF-2011-011*
[5] The ATLAS Collaboration 2011 *ATLAS-CONF-2011-035*
[6] S Frixione and BR Webber 2002 *JHEP* **0206** 029 (*Preprint* hep-ph/0204244)
[7] S Frixione and P Nason and BR Webber 2003 *JHEP* **0308** 007 (*Preprint* hep-ph/0305252)
[8] G Corcella *et al.* 2001 *JHEP* **0101** 010 (*Preprint* hep-ph/0011363)
[9] ML Mangano *et al.* 2003 *JHEP* **0307** 001 (*Preprint* hep-ex/0206293)
[10] M Cacciari, GP Salam and G Soyez 2008 *JHEP* **04** 63 (*Preprint* arxiv:0802.1189)
[11] M Cacciari, GP Salam and G Soyez The fastjet package http://fastjet.fr/
[12] S Frixione, P Nason and C Oleari 2007 *JHEP* **0711** 070 (*Preprint* arXiv:0709.2092)
[13] T Sjostrand and S Mrenna and PZ Skands 2006 *JHEP* **05** 026 (*Preprint* hep-ph/0603175)
[14] BP Kersevan and E Richter-Wąs 2003 *Comput. Phys. Commun.* **149** 142 (*Preprint* hep-ph/0201302)
[15] BP Kersevan and E Richter-Wąs 2004 (*Preprint* hep-ph/0405247)
[16] J Pumplin *et al.* 2002 *JHEP* **07** 012 (*Preprint* hep-ph/0201195)
[17] The CDF Collaboration A Abulencia *et al.* 2006 *Phys. Rev.* **D73** 032003 (*Preprint* hep-ex/0510048v1)
[18] The ATLAS Collaboration 2010 Prospects for the measurement of the top-quark mass using the template method with early atlas data ATL-PHYS-PUB-2010-004

Top and top-pair mass measurement at CMS

Roberto Chierici

CNRS, Institut de Physique Nucleaire de Lyon, 4 Rue Fermi, 69100 Villeurbanne

E-mail: `Roberto.Chierici@cern.ch`

Abstract. In this article the measurements of the top mass and the top-pair mass by using $t\bar{t}$ events produced in proton-proton collisions at 7 TeV with the CMS detector at the LHC [1] are presented. Particular emphasis will be given to the reconstruction challenges involving jets in the final state. The presented results correspond to the data collected by CMS in 2010 and an integrated luminosity of 36/pb.

1. Introduction

Top physics is one of the pillars of the physics program at the LHC. The top quark is the heaviest elementary particle known and is strongly coupled with the electroweak symmetry breaking sector. Indeed, the top mass is a fundamental parameter of the Standard Model (SM) whose precise knowledge greatly helps in constraining the model itself and in determining indirect constraint on unknown parameters, like for instance the Higgs boson mass. Moreover, top pair production is the perfect window for finding evidence of new physics: several models even predict favourable couplings of new particles to the third family.

Doing top physics at a hadron collider requires a throughout understanding of the detector and complex analysis tools. Depending on the decay of the W bosons, top pair production always leads to jets in the final state, and may also involve the production of charged leptons and neutrinos, originating missing transverse energy. The understanding of the hadronic component is therefore crucial for a full event reconstruction: in the following we will try to underline the aspects which are more important for a measurement, namely the impact of the knowledge of the jet energy scale, the jet pairing, heavy flavour tagging, and top tagging for boosted objects.

2. The top mass in the dilepton channel

The first measurement of the top mass at the LHC has been performed by CMS by using the di-lepton final states. In spite of a lower branching ratios and a more difficult reconstruction procedure, due to the presence of two neutrinos in the final state, the di-lepton channel is much more background free and the event selection straightforward. The event is required to contain two isolated, prompt, opposite charge leptons with p_T >20GeV/c and pseudorapidity $|\eta|$ <2.5. For same flavour leptons a mass window of 15 GeV/c^2 around the Z mass is vetoed. Jets are built from the individual particles reconstructed in the detector, matching the information from all the subdetectors CMS is composed of [2]. At least two jets with p_T >30 in the central detector region are required, and a cut on the missing transverse energy at 20 GeV (30 for same flavour lepton final states) is imposed. With a statistical sample of 36/pb, a total of 102 events have been selected, to be compared to 108.5 expected from MC, with an expected purity of about 90%.

For determining the top mass it is necessary to determine the neutrino momenta. The system is underconstrained but it can be solved by imposing the W and top masses in the event, once the lepton-jet assignment is chosen. In the pairing, jets which are b-tagged are favoured as candidates from the top decays. This is realised by two different methods, as described in [3], where the best hypothesis for the top mass is obtained by solving iteratively the equation for different values of a probe top mass. From the distribution of the best top mass hypotheses the value of m_t is then extracted by two template fits, as shown in figure 1. The signal is taken

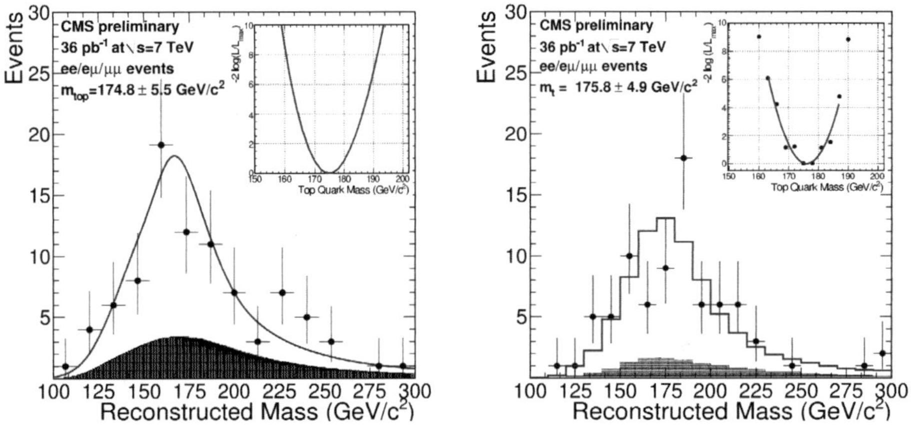

Figure 1. Reconstructed top masses in the two different methods of reference [3], and best template fits superimposed.

from the MC predictions at different values of the top mass, and the backgrounds are partially taken from data (in the case of Z+jets background) and from MC (Single top, top-pair, W+jets and di-bosons). The methods, individually calibrated on MC, show a linear behaviour on the top mass and a residual small bias, which is corrected for. The two analyses provide compatible results and a statistical correlation, studied via pseudo-experiments, of 0.57. They are combined and yield a top mass of:

$$m_t = (175.5 \pm 4.6(\text{stat}) \pm 4.6(\text{syst}))\text{GeV}/c^2 \quad (1)$$

where the dominant systematic error is the knowledge of the jet energy scale, assessed in CMS to be in a range between 2 and 4%, as a function of the jet energy and rapidity [4].

3. The top pair mass

Studying the top pair mass is particularly important as a measure in the frame of the SM, but also as a probe of the presence of new physics in the top sector. Several models [5] predict heavy s-channel resonances in top-pair production, that may give rise to resonant structures, or more generally a distortion in the shape of m(tt). The kinematics of top-pair production can be strongly influenced by the presence of new heavy particles. In particular, they may give rise to highly boosted top quarks, whose decay products may progressively merge, imposing new constraints in the way the events must be (triggered and) reconstructed. In what follows we present a full analysis optimized for the threshold of top-pair production, and studies for the reconstruction of highly boosted top pair quarks.

3.1. Top pair mass at the threshold of top pair production

This measurement is performed by exploiting semi-leptonic final states with the presence of an electron or a muon. Events are triggered by the presence of an isolated lepton, which is required to have a minimum transverse momentum of 20 GeV/c for muons and 30 GeV/c for electrons. A cut at 20 GeV on the minimum amount of missing transverse energy is required to substantially reduce the QCD component, and at least three(four) central jets are required with different thresholds in p_T of 70/50/30(/30) GeV/c. The selected events are divided in eight categories, four per lepton family, according to the number of jets and b-tags in the event. The four categories are composed by events with three jets with one b-tag, and four jets with either 0, 1 or 2 b-tags. The yields in data agree within errors with the predictions in all the eight categories, and sum up to 769 events seen in data with 822 expected. The expected number of events is inferred from MC with the exception of the QCD component, determined from control regions in data themselves.

The events can be reconstructed with a kinematic fit, where W and top masses are imposed. The main ambiguity in this process is the correct pairing of jets in forming the candidate top quarks. For this, a χ^2 method is used, choosing the pairing that gives the best similarity to a top-pair event for what concerns the leptonic and hadronic masses, the transverse momentum of the top-pair system, and the scalar sum of the transverse energy in the event. It has been verified that this procedure introduces no significant bias in the search of new physics in the top-pair production.

The improvement of the use of a kinematic fit is shown in figure 2, which illustrates

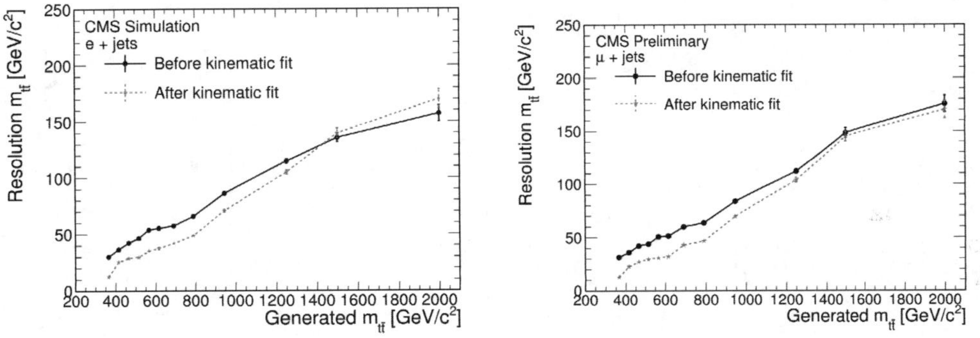

Figure 2. Expected resolution on the mass of the top mass pair before and after a kinematic fit, for the semileptonic muon channel and the electron one.

the reduction of the mass resolution when imploying a fit to the event.The improvement is significative up to invariant masses of about 1.5 TeV/c^2. For events with only three jets, no kinematic fit is applied.

The mass distributions seen in data agree well with the expectations in all the categories, as detailed in [6]. As an illustration, the distributions of the four jet bins with one and two b-tags in the muon channel are reported in figure 3 with the full 2010 statistics. The MC distributions have been normalized to the collected luminosity. Since no excess is found, the observed distributions in data are used to set limits on a new potential narrow resonance in top-pair production. The statistical approach for doing this is a fully Bayesian one, where the mass distributions are modelled by templates for the signal (a generic Z' narrow-width resonance with SM couplings to fermions) and the backgrounds (mostly for MC, except for QCD, taken from data sidebands). The templates are parametric in all uncertainties, which are included as nuisance parameters to be integrated out. The parametric dependence on nuisances in each

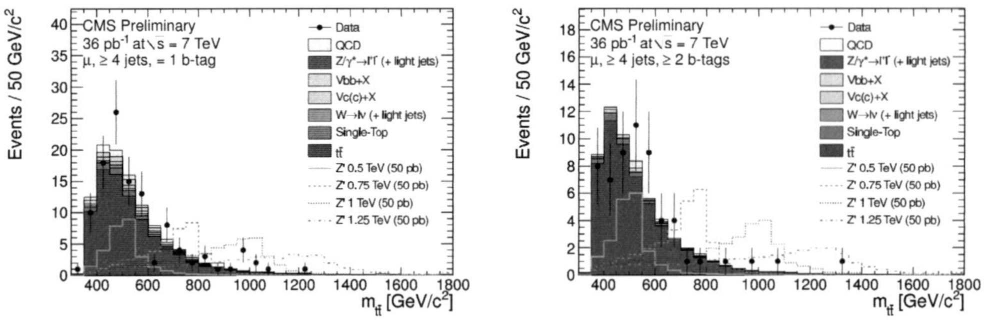

Figure 3. Top pair mass distributions for the muon channel in the four jets, one and two-b-tags categories.

bin is fitted via cubic functions, so that their impact in both rates and shapes can be taken into account. The most important nuisance parameters from systematic sources are background uncertainties, including theory ones like the Q^2 dependence, jet energy scale and resolution uncertainties, b-tagging efficiency and luminosity.

An integration of the likelihood over nuisances is performed in order to derive upper limits on the production cross section of a narrow resonance. They are shown in figure 4, as a function

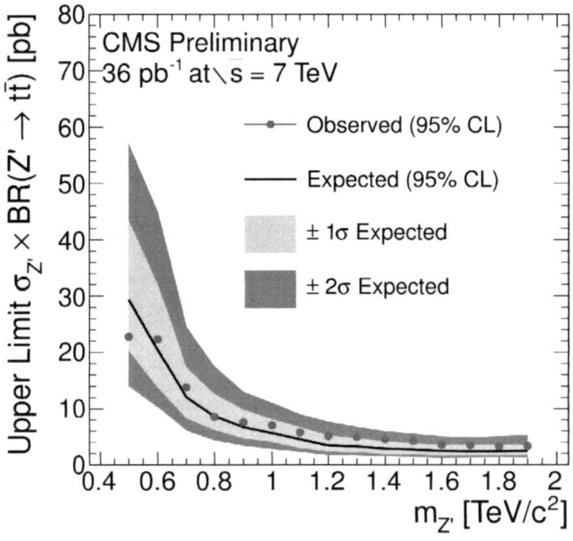

Figure 4. Expected and observed 95% upper limit on the production cross section of a generic Z'. The band includes statistical and systematic uncertainties.

of the resonance mass. The figure shows both the expected limit, and the one observed in data. Both are in agreement, with no visible excess in the entire mass range in the reach of this analysis. With only 36/pb of data, models predicting cross sections of about 10 pb for masses of about 1TeV can already be excluded.

3.2. Boosted tops

As discussed, boosted top configurations may originate from the decay of a very massive object. In these circumstances leptons from W decays may progressively loose their isolation and the jets from the hadronic top deays overlap partially or totally, making it impossible to recur to standard reconstruction techniques.

In such final state topologies it becomes essential to be able to analyse jet substructures. The method investigated in CMS, called top tagging, uses a modified Cambridge-Aachen (C-A) algorithm, as described in [7] and in [8]. Jets are first clustered using the C-A algorithm, which does not use any k_T weighting in the pairing, with a large value of R (0.8). The found jets are decomposed by looking two steps back in the pairing history, searching for what we may call the "parents" and "grandparents" jets which have originated the final one. If the two parents have each at least 5% of the energy of the resulting jet, and at least the grandparents corresponding to one of the two parents have the same property, then the decomposition is declared successful. In these cases the jet is therefore decomposed in either three or four subjets. To be able to disentangle boosted top jets with respect to QCD, cuts on these substructures can be imposed. The two main variables that are used are m_{jet}, the mass of the original jet, and m_{min}, the minimum invariant mass of all the subjet pairs composing the jet. These variables are shown, for an hypothetical Z' signal into top-pair and for QCD events, as a function of the transverse momentum of the originary jet, in figure 5. As can be appreciated from the right hand side

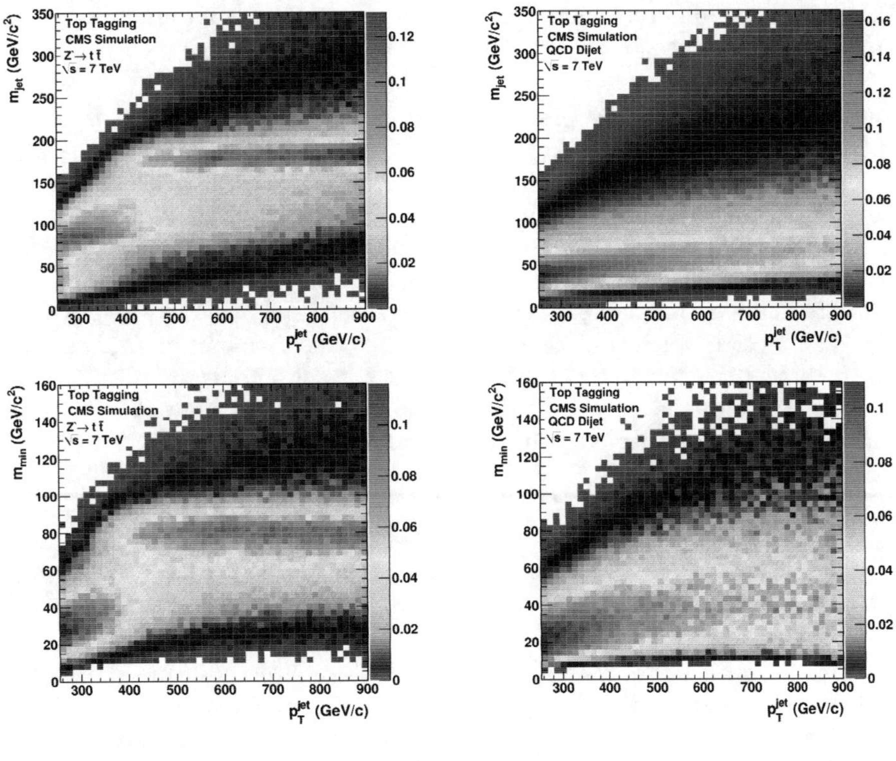

Figure 5. Jet mass (upper row) and minimum di-jet mass of all jet pairs composing the original jet (lower row) as a function of the jet p_T for a Z' signal into top pairs (left column) and QCD events (right column).

plots in the figure, as the originary jet gets boosted, the more likely it is that it collectes all the decay products of the top quark. The merging of the W boson is also visible in the plots.

No such structures are present for pure QCD jets. The cuts imposed on these variables are 100 GeV/c^2 <m_{jet} <250 GeV/c^2 and m_{min} >50 GeV/c^2. We refer to this procedure as top tagging. The efficiency of top tagging is determined on Z' signal samples as a function of the boost of the originary jet. Efficiencies of about 40-50% for boosts between 600 GeV to 1 TeV are found, going down to about 30% for 2 TeV or more.

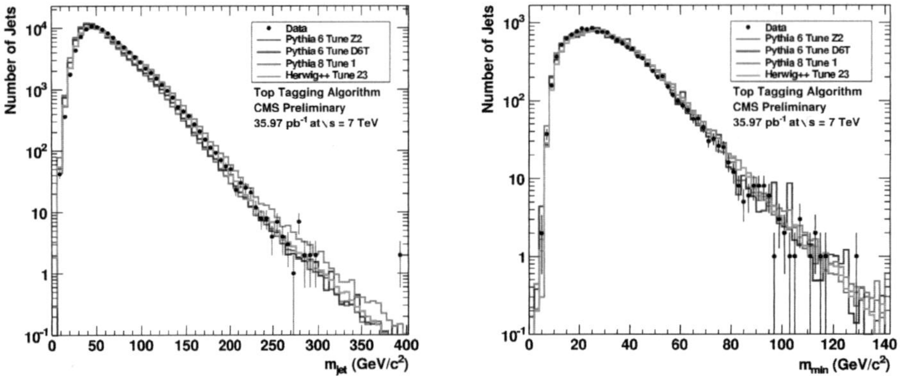

Figure 6. Predictions, for different models, of the jet mass and the minimum di-jet mass of all jet pairs composing the original jet for QCD events

Both m_{jet} and m_{min} are already pretty well described in pure SM QCD production by the current hadronisation models, as shown in figure 6. The dependence of the shape of the histograms on the different tunings is mild, increasing the robustness of the choice of these variables as good discriminators between boosted top signal and QCD background.

An essential ingredient for an analysis based on top tagging is to have precise knowledge of the QCD background. This can be determined in a simple way directly from the data by using di-jet events where one side is anti-top tagged and by parametrizing the tagging efficiency in the probe side as a function, for instance, of the jet transverse momentum. This is illustrated in figure 7 for 2010 data. The top mistag rate is shown as a function of the jet p_T. Once again, the level of agreement between what is observed from the data and what predicted by the various parton shower and hadronization models in the MC is remarkable.

4. Conclusions

Studying top pair production at the LHC is fundamental for both constraining the SM and searching for hints of presence of new phisics at the TeV scale. Two important analyses in this respect involve the determination of the top mass and the reconstruction of the mass of the top-pair. Both tasks need an excellent understanding of jet production and pairing, and jet energy calibration. Techniques to improve in all aspects are already in place in both analyses. The searches for new physics signals in top-pair production will also entail new issues in jet reconstruction, namely the ability of resolving jet substructures. This is also being successfully addressed in CMS.

[1] The CMS Collaboration, JINST, 3:S08004, 2008.
[2] The CMS Collaboration, CMS-PAS-PFT-10-002, 2011.
[3] The CMS Collaboration, CMS-PAS-TOP-11-002, 2011.
[4] The CMS Collaboration, CMS-PAS-JES-10-010, 2010.
[5] S. Dimopoulos and H. Georgi, Nucl. Phys. B193, 150
S. Weinberg, Phys. Rev. D13, 974

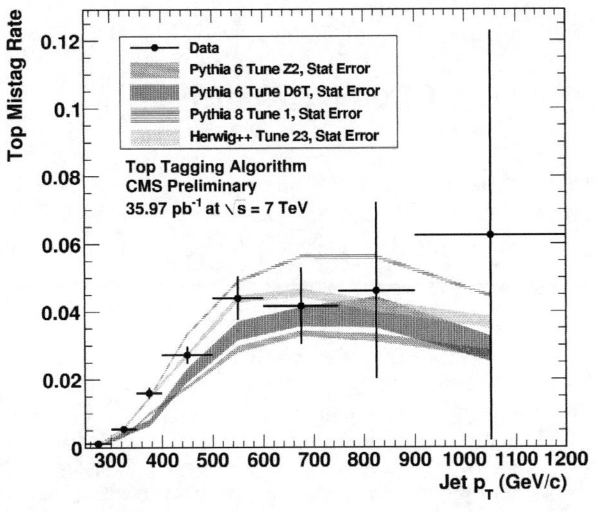

Figure 7. Top mistag rate as determined fron 2010 QCD data.

L. Susskind, Phys. Rev. D20, 2619
C. Hill, J. Parke, Phys. Rev. D49, 4454
R. Chivukula et al, Phys. Rev. D59, 075003
N. Arkani-Hamed, A. Cohen and H. Georgi, Phys. Lett. B513, 232
N. Arkani-Hamed et al, Phys. Lett. B429, 263
L. Randall, R. Sundrum, Phys. Rev. Lett. 83, 3370.
[6] The CMS Collaboration, CMS-PAS-TOP-10-007, 2011.
[7] D. Kaplan et al., Phys. Rev. Lett. 101, 142001.
J. Ellis et al., Phy. Rev. D80 (2009)
[8] The CMS Collaboration, CMS-PAS-JME-10-013, 2010.

W/Z+jets with the CMS detector

Massimo Nespolo
on behalf of the CMS collaboration
Universitá di Padova and INFN, Via Marzolo 8, 35131 Padova, Italy
E-mail: massimo.nespolo@gmail.com

Abstract. We present a study of jet production in association with W and Z bosons in proton-proton collisions at $\sqrt{s} = 7$ TeV using the full 2010 data set collected by the CMS experiment corresponding to an integrated luminosity of 36±4 pb^{-1}. The production of vector bosons with jets provides a stringent and important test of perturbative QCD, and is an important background to the study of top quark properties and in searches for new physics. We report the inclusive rates of jets produced (normalized to the total cross section) as well as the ratios $\sigma(V + (n+1)\text{-jets})/\sigma(V + n\text{-jets})$ for a jet threshold of 30 GeV. We also present a measurement of $Z \to ll + b$ and the first test of the Berends-Giele scaling at $\sqrt{s} = 7$ TeV.

1. Motivation

The production of vector bosons with jets provides a stringent test of perturbative QCD, and is an important background to the study of top quark properties and in searches for new physics. Events with jets produced in association with W and Z bosons are usually simulated with Matrix Element + Particle Showers generators (MadGraph[1]+PYTHIA[2]). This means that the description of this process is at the tree level, so the overall normalization is uncertain, and a procedure to combine partons and fragmentation is needed. This situation has improved due to next-to-leading order (NLO) calculations, which were recently performed for up to W+4 jets [3], and up to Z+3 jets [4]. However, also in these cases the precision varies from 10% to 30%, due to uncertainties on parton distribution function, and due to the choice of renormalization and factorization scale.

It is therefore essential to use events tagged by the presence of W/Z bosons at the hadron colliders to probe QCD calculations in an environment which is relatively clean. In this way, it is possible to extract background normalizations for lepton(s)+MET+jets final states, which are important not only for Standard Model (SM) measurements (top, Higgs), but also for searches beyond the Standard Model (BSM), such as SUSY.

Moreover, we can observe that this analysis in itself may also be potentially sensitive to new physics, since heavy particles predicted by several BSM theories can show their presence through an excess of jets at high p_T, or through a deviation from the expected scaling.

2. CMS Detector

The central feature of the CMS detector is a 6 m diameter superconducting solenoid which provides a 3.8 T axial field. Located within the field volume are the silicon pixel and strip tracker, the crystal electromagnetic calorimeter and the brass/scintillator hadron calorimeter.

Muons are measured in gas-ionisation detectors embedded in the steel return yoke. A detailed description of the detector and its performance can be found in [5].

3. Selection and First Results

For the muon channel, we only use events which, according to the High Level Trigger (HLT) system of the CMS detector, have at least one muon with $p_T >$ (9-15) GeV (the threshold was raised following the increasing instantaneous luminosity of the LHC). In offline analysis, we require presence of one muon well fitted from hits in the tracker and in the muon chambers and consistent with ID designed to reject punch-through and decay in flight. This muon must have $p_T > 20$ GeV, $|\eta| < 2.1$, and must be isolated: $(\sum p_T(tk) + \sum E_T(had + em))/p_{T\mu} < 15\%$. If there is a second muon with $p_T > 10$ GeV, $\eta < 2.4$ and 50 GeV $< M_{\mu\mu} <$ 120 GeV, the event is put in the Z sample; otherwise, if $M_T(\mu\nu) > 20$ GeV, it is put in the W sample.

Similarly, in the electron channel we select only events with one electron with $p_T >$ (10-17) GeV at the HLT. For the final analysis, we require presence of one electron satisfying **tight** (80% efficient) isolation, electron identification, and conversion rejection criteria. This electron must have $p_T > 20$ GeV, $|\eta| < 2.5$ ($1.44 < |\eta| < 1.57$ excluded), and must match the trigger primitive. Events were assigned to the W or to the Z sample using the same criteria described for the muon channel, with the only difference that the second electron was required to pass a **loose** (95% efficient) ID, and that the η cut was identical for the two electrons.

Jets are reconstructed using the *anti-kt* clustering algorithm [6] with $\Delta R = 0.5$ (the default in CMS), applied to a list of "particles" identified by the Particle Flow framework [7] (PF-jets). These jets are accepted only if they have $E_T > 30$ GeV and $|\eta| < 2.4$ (which is the tracker acceptance). We also apply a loose ID to remove calorimetric noise [8], apply data-driven energy calibration, and pile-up energy offset removal procedure based on the event energy density (see [9] and [10]). Leptons from Vector Boson decay should not be counted as jets, so isolated muons are removed from the list of particles before PF jets are clustered, and jets which have $\Delta R < 0.3$ from the selected electron are also not considered. Additional details are given in [11].

The transverse momentum distribution of the leading jet for W+1 jet and Z+1 jet is shown in figures 1-4, while the distribution for exclusive jet multiplicity in W/Z+n jets (n=0 to 6) is shown in figures 5-8. To remove further background, an additional cut of $M_T > 50$ GeV is applied to the W events. The data is in very good agreement with the MadGraph predictions normalized to the NNLO cross sections.

4. Signal Extraction and Corrections

The MC simulations show that the procedure described in the previous section to identify the Z boson produces a very clean resonant peak with extremely low background for all jet multiplicity; the corresponding signal is therefore extracted with an Unbinned Maximum-Likehood (UML) fit to the di-lepton invariant mass spectra. In this case, a functional form is assumed (Cruijff for signal, exponential accounting for all backgrounds) with floating background and the signal parameters also are left floating, but are kept equal for all jet multiplicity.

The situation is more complicated for W events, because its experimental signature (a single isolated high-p_T lepton with missing transverse energy) can be mimicked by many processes for physical reason (i.e. presence of a real W) or for detector effects (mainly acceptance and fake leptons). The residual background in the transverse mass spectra can be divided into two categories: not peaking at M_W (mainly QCD), and peaking at M_W (top, which is important in W+3,4 jets). To cope with this, an UML 2-dimensional fit is performed on the M_T distribution vs the number of b-jets $n_{jets}^{b-tagged}$, the latter being identified with the "Track Counting High Purity" b-tagging algorithm for the muon channel and with the "Track Counting High Efficiency" algorithm for the electron channel [12]. Also in this case a functional form is assumed, and the fit is constrained in the range $20 < M_T < 150$ GeV. Two parameters of the fit (b-tagging

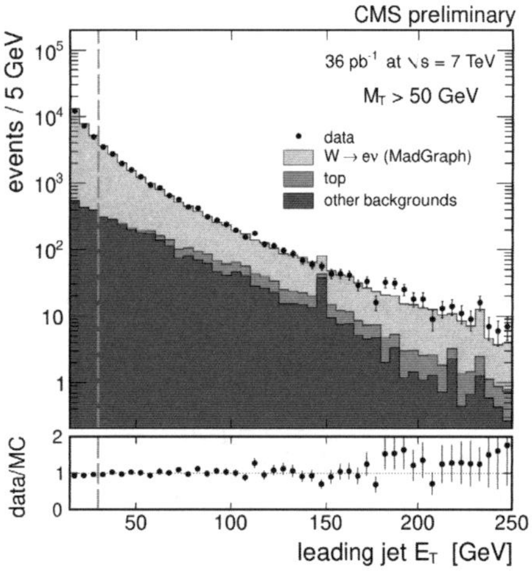

Figure 1. The p_T distribution of the leading jet in the W+1 jet sample (electron channel), compared with MC.

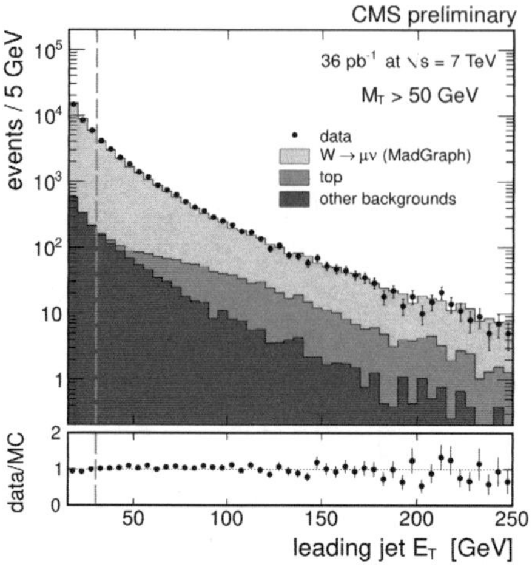

Figure 2. The p_T distribution of the leading jet in the W+1 jet sample (muon channel), compared with MC.

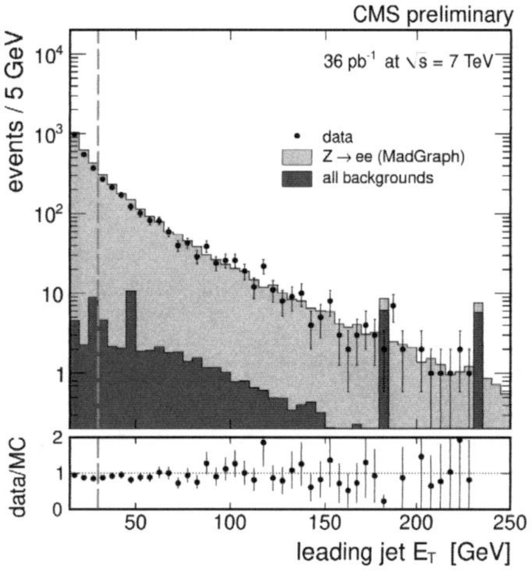

Figure 3. The p_T distribution of the leading jet in the Z+1 jet sample (electron channel), compared with MC.

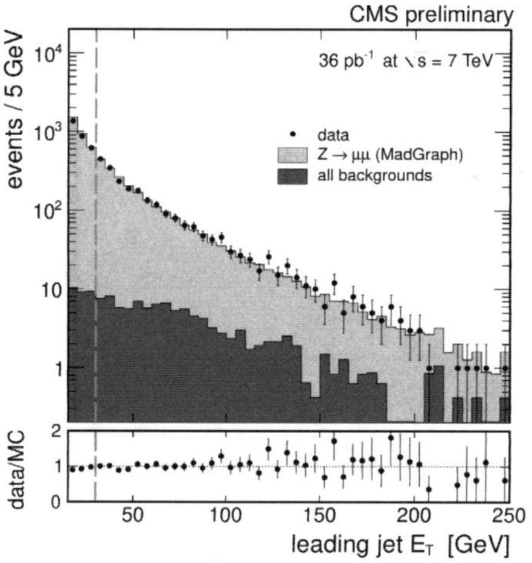

Figure 4. The p_T distribution of the leading jet in the Z+1 jet sample (muon channel), compared with MC.

efficiency and mis-tag probability) are fixed to the corresponding values extracted from MC, and are cross-checked with a di-leptonic top-enriched data sample.

Many corrections had to be applied to the fitted yields: first of all, we took into account the lepton efficiency measured with *Tag & Probe* method on Z events, and factorized into reconstruction, selection, and trigger efficiency. For the W candidates, we corrected (using MC and data) for the fact that the range $0 < M_T < 20$ GeV was not included in the fit. Finally, to remove the effects of imperfect energy resolution and reconstruction efficiency, we used MadGraph simulation to extract the migration matrix $\mathrm{R}(n_{RECO}, n_{Gen})$, and applied a Singular Value Decomposition (SVD, see [13]) to "unsmear" the n_{jet} distribution.

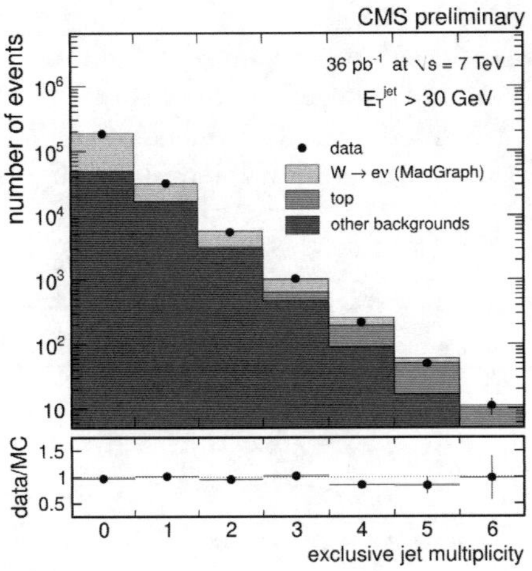

Figure 5. Exclusive reconstructed jet multiplicity with $W \to e\nu$, not corrected for efficiency and detector effects.

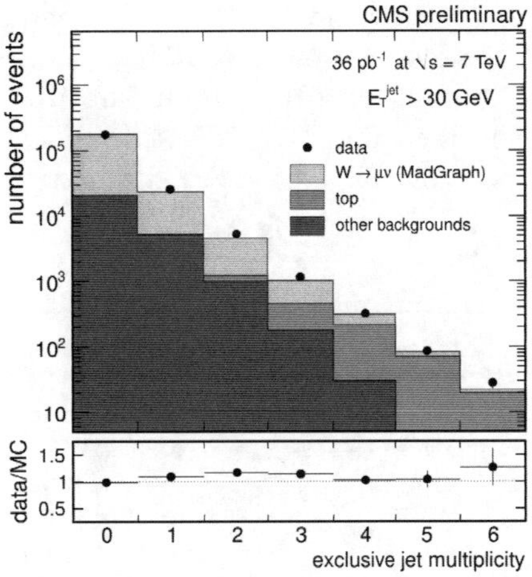

Figure 6. Exclusive reconstructed jet multiplicity with $W \to \mu\nu$, not corrected for efficiency and detector effects.

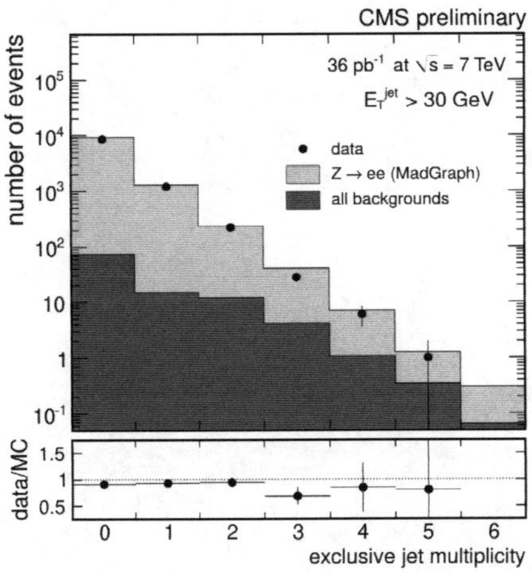

Figure 7. Exclusive reconstructed jet multiplicity with $Z \to e^+e^-$, not corrected for efficiency and detector effects.

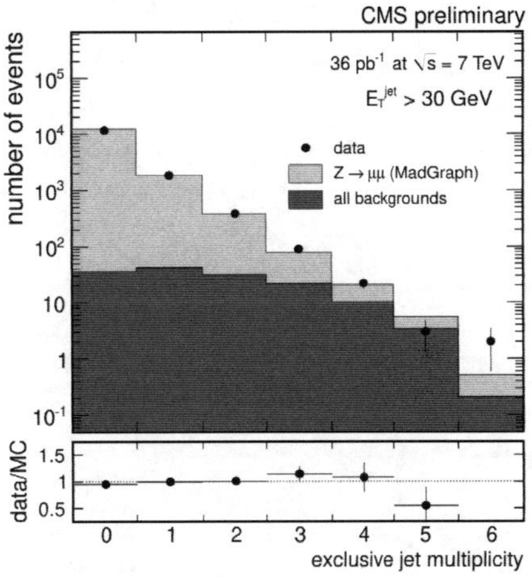

Figure 8. Exclusive reconstructed jet multiplicity with $Z \to \mu^+\mu^-$, not corrected for efficiency and detector effects.

5. Ratios, Berends-Giele Scaling and Systematics

Finally we measured two types of ratios: the production of $\geq n$ jets over the total cross-section, and the ratio of $(V+ \geq n\text{-jets})/(V+ \geq (n-1)\text{-jets})$. The usage of ratios greatly reduces the systematics. We obtain in fact a full cancellation of the contribution due to luminosity (which had a big uncertainty, approximately 10% at the time this measurement was done) and a partial cancellation of the effects produced by the jet energy scale, so only lepton efficiency versus the number of jets matters. The main source of the remaining systematic uncertainty is the inaccuracy in the determination of the jet energy which moves jets above and below certain threshold, changing the measured jet multiplicity. This effect is produced by the uncertainties on data-driven JEC, by the JEC flavour dependence (estimated from MC), and by pile-up removal (500 MeV in excess for each jet in MC). Moreover, we also considered jet energy resolution, the pile-up residual effect after pile-up subtraction, and for the W, the effect of mismeasurement on MET (from fitting M_T on data). Finally, the effect of reconstruction/selection efficiency (close to the previous one) and signal extraction (important only at high jet multiplicity) were evaluated.

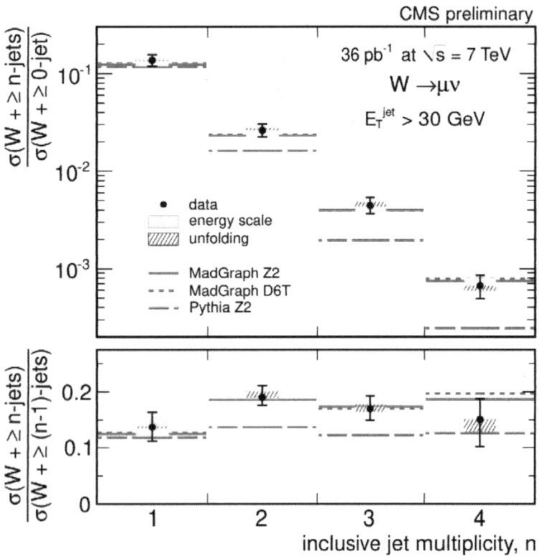

Figure 9. Ratios for $W \to \mu\nu$, compared with expectation from MadGraph and PYTHIA.

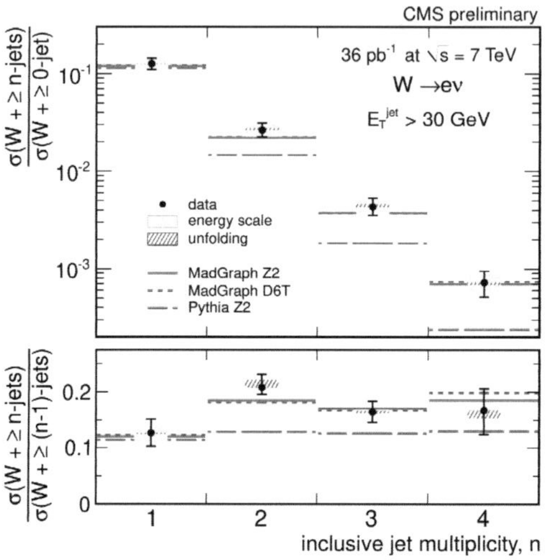

Figure 10. Ratios for $W \to e\nu$, compared with expectation from MadGraph and PYTHIA.

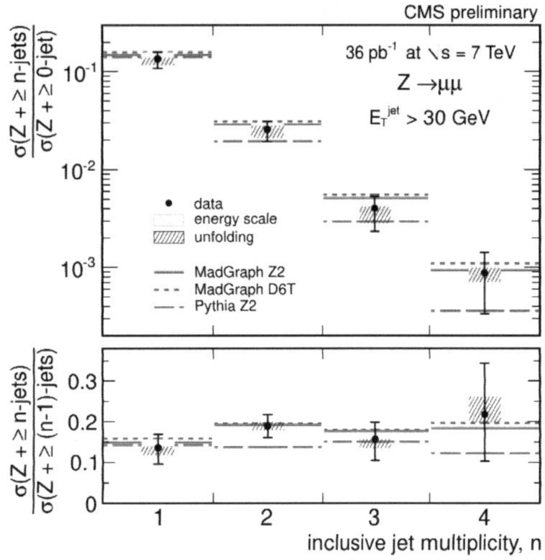

Figure 11. Ratios for $Z \to e^+e^-$, compared with expectation from MadGraph and PYTHIA.

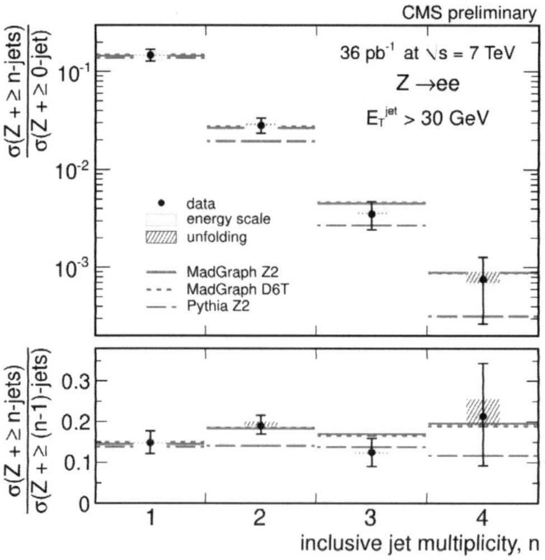

Figure 12. Ratios for $Z \to \mu^+\mu^-$, compared with expectation from MadGraph and PYTHIA.

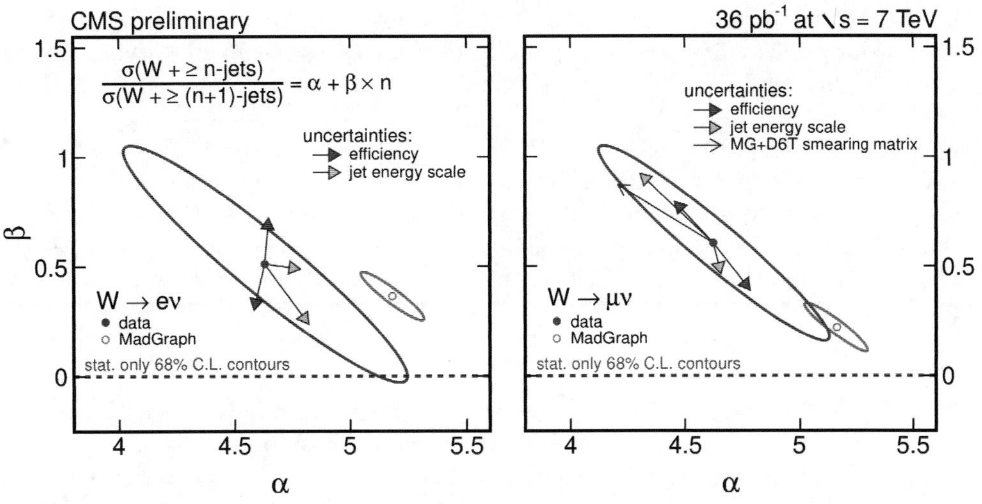

Figure 13. Fit results for W+jets using Berends-Giele scaling parameters after pile-up subtraction, efficiency corrections, and unfolding for detector resolution effects. The data are compared to MadGraph with Z2 tune.

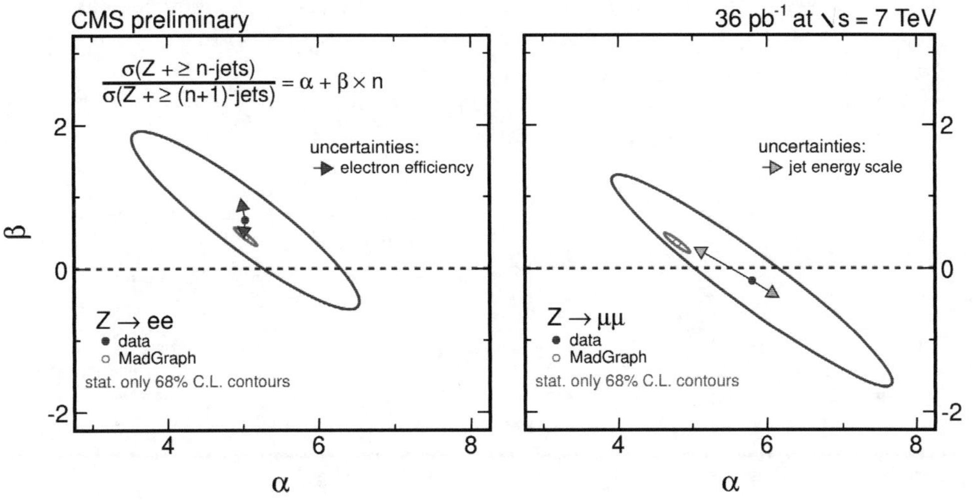

Figure 14. Fit results for Z+jets using Berends-Giele scaling parameters after pile-up subtraction, efficiency corrections, and unfolding for detector resolution effects. The data are compared to MadGraph with Z2 tune.

The results are shown in figures 9-10 for the W+jets (muon and electron channel respectively), and in figures 11-12 for the Z+jets (again, for the muon and electron channel respectively). The measured ratios are in excellent agreement with MC predictions from ME+PS (MadGraph), while PYTHIA alone start to fail for $n_{jets} \geq 2$ in W+jet events. In Z+jet events both calculations are compatible with data because of limited statistics and hence larger errors.

For what concerns the Berends-Giele scaling [14], the LO calculation would predict a constant value for the ratio $C_n = \sigma_n/\sigma_{n+1}$, but the NLO corrections and/or phase space effects could violate this simple proportionality. To check possible deviations, we assumed a linear dependence from n ($C_n = \alpha + \beta n$), and we performed a 2-parameter fit on the *exclusive* jet multiplicity spectra.

The results obtained (excluding events with no jet recoiling against the Vector Boson, and taking into account the bin-to-bin migration) for W and Z are shown in figure 13 and 14 respectively. In the first case, β is compatible with 0 within 1σ, and the difference between the two decay channels are due to the ΔR cut (which is not applied to the muons); the Z plots show that β is even more compatible with 0 (within 0.5σ), and that there is a very good agreement with the ME+PS prediction.

6. Z + *b* jets

The production of Z decaying into leptons in association with b-quarks is a benchmark channel for the identification of Higgs (especially MSSM Higgs) at the LHC, and is also a background for SM and BSM searches involving leptons + b-jet final states. However, theoretical calculations of the corresponding cross section cannot be precise enough, since there is a 30% difference ([15], p.53) between the numbers obtained with the fixed-flavour scheme (where b-quarks are produced explicitly in pairs from gluon splitting), and those predicted using the variable-flavour scheme (which allows the b-quarks to participate directly in the hard scattering, by integrating the gluon splitting process into the PDF).

For this measurement (see [16]), events were selected requiring one Z (decaying either into muons or electrons) and at least one PF jet with $E_T > 25$ GeV, $|\eta| < 2.1$ (which implied that the jet was contained completely inside the tracker), ΔR(jet, lepton) > 0.5, and one or more secondary vertices reconstructed inside the jet itself. Two versions of b-tagging called Simple Secondary Vertex (SSV, see [12]) with high efficiency and high purity were used, and top candidates were rejected by applying the cuts MET > 40 GeV, and 60 GeV $< M_{ll} <$ 120 GeV.

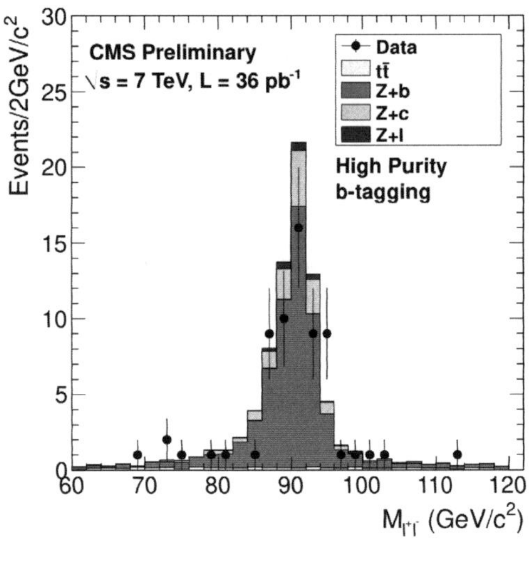

Figure 15. Invariant mass of opposite-sign lepton pairs for the high-purity selection, compared with MC simulation.

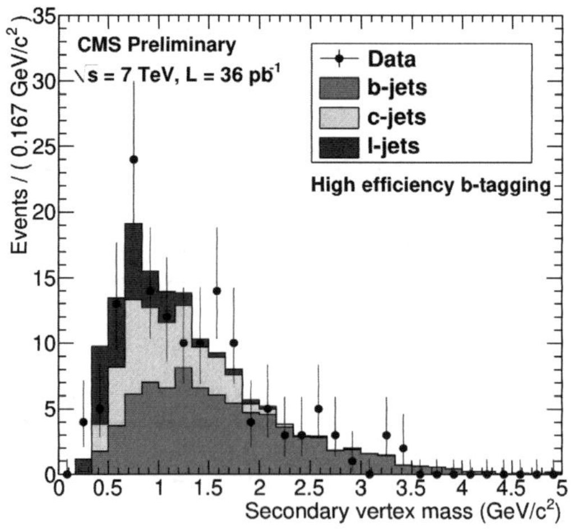

Figure 16. Binned likelihood fit of the mass of the secondary vertex in data events, for the high-efficiency selection, compared with MC simulation.

The resulting invariant mass spectrum for the High-Purity SSV b-tagging algorithm is shown in figure 15, and has been compared with the MC expectation for Z + different quark flavour. The purity of this sample was extracted with a binned likehood fit of the mass of the secondary vertex (see figure 16 for the High-Efficiency SSV), using templates from MC. Both the measured purity and the ratio $(Z+b)/(Z+j)$ are in good agreement with MadGraph simulation within theoretical and experimental uncertainties.

7. Conclusions

We have studied the production of jets in association with a W or a Z boson, and the production of b-jets in association with a Z, where the Vector Boson decays into electrons or muons with the full dataset of 36±4 pb^{-1} collected by the CMS detector in 2010.

We measured the leading jet p_T spectra, the ratios $\sigma(V+n\text{-jets})/\sigma(V)$ and $\sigma(V+n\text{-jets})/\sigma(V+(n-1)\text{-jets})$, the Berends-Giele scaling, and the ratio $(Z+b)/(Z+j)$. In all cases, the agreement with MC calculation (ME+PS, in particular) is very good, and the precision achieved shows that CMS is ready for searches beyond the Standard Model.

References

[1] F. Maltoni and T. Stelzer, "MadEvent: Automatic event generation with MadGraph", *JHEP* **02** (2003) 027, *arXiv:hep-ph/0208156.*

[2] T. Sjostrand, S. Mrenna, and P. Z. Skands, "PYTHIA 6.4 Physics and Manual", *JHEP* **05** (2006) 026, *arXiv:hep-ph/0603175.*

[3] C. F. Berger, Z. Bern, L. J. Dixon et al., "Precise Predictions for W + 4-Jet Production at the Large Hadron Collider", *Phys. Rev. Lett.* **106** (Mar, 2011) 092001.

[4] C. F. Berger, Z. Bern, L. J. Dixon et al., "Next-to-leading order QCD predictions for Z, γ^* + 3-jet distributions at the Tevatron", *Phys. Rev. D* **82** (Oct, 2010) 074002.

[5] CMS Collaboration, "The CMS experiment at the CERN LHC", *JINST* **3** (2008) S08004.

[6] M. Cacciari, G.P. Salam, and G. Soyez, "The anti-kt jet clustering algorithm", *JHEP* **0804** (2008) 063.

[7] CMS Collaboration, "Commissioning of the Particle-Flow Event Reconstruction using high-energy LHC Collisions in the CMS Detector", *CMS PAS* **PFT 1-002** (2010).

[8] CMS Collaboration, "Identification and Filtering of Uncharacteristic Noise in the CMS Hadron Calorimeter", *JINST* **5** (2010) T03014, *arXiv:0911.4881.*

[9] M. Cacciari and G.P. Salam, "Pileup subtraction using jet areas (hep-ph/0707.1378)", *Physics Letters B* **659** (2008) 119126.

[10] M. Cacciari, G.P. Salam and G. Soyez, "The catchment area of jets (hep-ph/0802.1188)", *JHEP* **0804** (2008) 005.

[11] The CMS Collaboration, "Rates of Jets Produced in Association with W and Z Bosons", *CMS PAS* **EWK 10-012** (2011).

[12] CMS Collaboration, "Commissioning of b-jet identification with pp collisions at $\sqrt{s}$ = 7 TeV", *CMS PAS* **BTV 10-001** (2010).

[13] A. Hocker and V. Kartvelishvili, "SVD Approach to Data Unfolding", *Nucl. Instrum. Meth. A* **372** (1996) 469481, *arXiv:hep-ph/9509307.*

[14] F. A. Berends, W. T. Giele, H. Kuijf et al., "Multi-jet production in W, Z events at $p\bar{p}$ colliders, *Phys. Lett. B* **224** (1989) 237.

[15] LHC Higgs Cross SectionWorking Group, "Handbook of LHC Higgs Cross Sections: 1. Inclusive Observables", *arXiv:hep-ph/1101.0593.*

[16] The CMS Collaboration, "Observation of $Z+b$, $Z \rightarrow ee$, $\mu\mu$ with CMS at $\sqrt{s}$ = 7 TeV", *CMS PAS* **EWK 10-015** (2011).

Search for diboson production in lepton+MET+$b\bar{b}$ channel

M Trovato[1] **and C Vernieri**[1]

[1]INFN-Pisa, Polo Fibonacci Largo B. Pontecorvo, 3 - 56127 Pisa, Italy

E-mail: `mtrovato@FNAL.GOV, cvernier@FNAL.GOV`

Abstract. The study of associated boson production in the final state with a lepton, neutrino, and a bb-pair is important since the event topology of this process is the same as expected for associated production of a W and the Standard Model light-Higgs boson. Thus, the search for boson production can be considered as a preliminary step towards the Higgs discovery. Here we present a search for WZ/ZZ in events with a lepton, missing transverse energy, and b-quark jets. Besides looking at the sample where two exclusive jets are found, we investigate the sample with 3 jets which contains 40% of the signal events. In this sample the invariant mass of the two E_T-leading jets would normally be chosen to signal the Z boson. We describe an alternative procedure to reconstruct the Z-invariant mass in this sample. This procedure exploits the information carried by the third jet in the sample.

1. Introduction

Measurements of diboson production at the Tevatron have been used to test of the electroweak sector of the Standard Model (SM) and to place limits on models of physics beyond the SM [?]. Diboson measurements are also instrumental for searches for the SM light Higgs boson at the Tevatron. By choosing to focus on the final state where a Z-boson decays into $b\bar{b}$-pairs, the event topology would be the same as expected for associated production of a W and a light Higgs boson ($m_H < 135\ GeV/c^2$). At the Tevatron, the process $WH \rightarrow Wb\bar{b}$ has an expected $\sigma \cdot BR$ [1] about five times lower than $WH \rightarrow Wb\bar{b}$ for $m_H \simeq 120\ GeV/c^2$. Therefore, observing this process would be a benchmark for the even more difficult search for the light Higgs in the $WH \rightarrow Wb\bar{b}$ process.

This paper describes a preliminary study towards improving the signal acceptance in the cross-section measurement of WZ and ZZ associated productions in a final-state with a lepton, missing transverse energy, and up to three jets, where two jets will be required to carry b-flavour ("b-tagged jets"). A first mandatory step of such a measurement is to validate how the simulations models the data on the aforementioned final-state sample, prior to any b-tagging requirements. This sample, named "pretag sample", benefits a larger statistics, and is therefore sensitive even to small mis-modelings. In Sec. 2 simulations will be compared to data in the pretag sample, where we apply the standard kinematical cut at CDF which requires the number of jets to be exclusively two. However, simulations show that only about half of the signal events pass this exclusive two-jets cut. In Sec. 3 we will investigate the sample with three jets in the final state. The problem of how best to reconstruct the correct Z-boson mass will be addressed.

[1] BR being the Branching ratios of $Z/H \rightarrow b\bar{b}$

2. Two jets pretag sample

In order to spot possible mis-modelings in our simulations the pretag sample is investigated.

2.1. Sample selection

We investigate the finale state in which a W or Z boson decays leptonically and the other Z-boson hadronically. The experimental signature involves the presence of a charged lepton (electron or muon) a neutrino (identified through the missing transverse energy, $E_{T,miss}$) and large-E_T jets. The jet E_T is corrected for known instrumental and physics effects. The sample we investigate is selected by applying the following cuts:

- exactly two jets with $E_T > 15$ GeV and $|\eta| < 2$
- an isolated triggered electron with $|\eta| < 1.1$ and E_T 20 GeV
- or muon with $|\eta| < 1.$ and $P_T > 20$ GeV
- $E_{T,miss} > 20$ GeV
- $E_{Tj1,j2} > 25$, 15 GeV and $|\eta_{j1,j2}| <2$; $j1$, $j2$ being the first and second leading E_T-jets
- $M_T^W > 10$ (30) GeV if the triggered lepton is a muon (electron), M_T^W being the W-invariant mass in the transverse plane
- $E_{T,miss}$-significance > 1.8 if the triggered lepton is an electron. The definition of $E_{T,miss}$-significance is given in [3]

2.2. SM processes composition of the selected sample

The following processes would contribute to a data sample selected sample within our cuts:

- **Electroweak and top (EW)**: WW, WZ, ZZ, Z+jets, $t\bar{t}$, single-*top*. Each of these processes can mimick the signal signature, with one detected lepton, large $E_{T,miss}$, and two jets. The contamination of these processes in the selected data sample is estimated by using their accurately predicted cross sections [4]. The shapes (templates) of a number of observables are obtained from ALPGEN+Pythia [5], [6], Pythia MC [6] after the simulation of the CDF detector.
- **W(→ lν)+jets,** $l = e, \mu, \tau$. Due to the presence of real leptons and neutrinos, the $W+jets$ background is the hardest to be reduced. Templates are obtained from ALPGEN+Pythia MC, while the rate normalization is obtained from data (see Fig.1)
- **QCD**: multi-jet production with a jet faking the lepton and fake $E_{T,miss}$. Since the mechanism for a jet faking a lepton or for fake missing transverse energy is not expected to be well modeled in MC events both rate normalization (see Fig.1) and templates are obtained from data. The way how ackground shapes are obtained for QCD events is explained elsewhere [7]

2.3. Shape validation

In this section we compare a number of significant observables to data and expectations (Fig.2, 3, 4) in order to validate the used MC, the QCD templates, and the expected rates in the pretag sample.

We see some mis-modeling of the data, in the lowest jet E_T bins. We are planning to investigate how other generators for W+jets events would model our data. Other QDC models are under investigation as well.

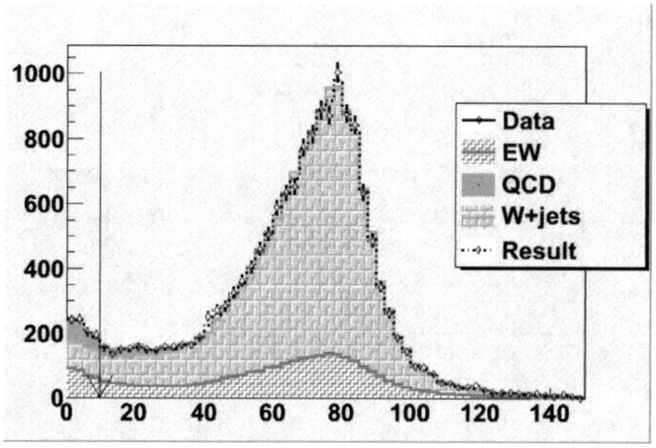

Figure 1. Fit on M_T^W for the QCD and W+jets rate in data collected with the high-P_T muon trigger. The dotted curve show the result of the fit.

Table 1. Predicted and observed number of events in the pretag sample. W+jets and QCD rates are estimated from the fitting data. The expected rates are separated for different triggered lepton type. By construction the expected number are equal to observed numbers. The calculation of uncertainties on these rates is still on-going.

Process	Rate (Electrons)	Rate (Muons)
Signal(WZ+ZZ)	125.7	120.8
WW	745.2	544.6
EW	1213.7	2062.9
$W + jets$	14874.9	12353.7
QCD	1937.0	216.8
Total Observed	18990	15497

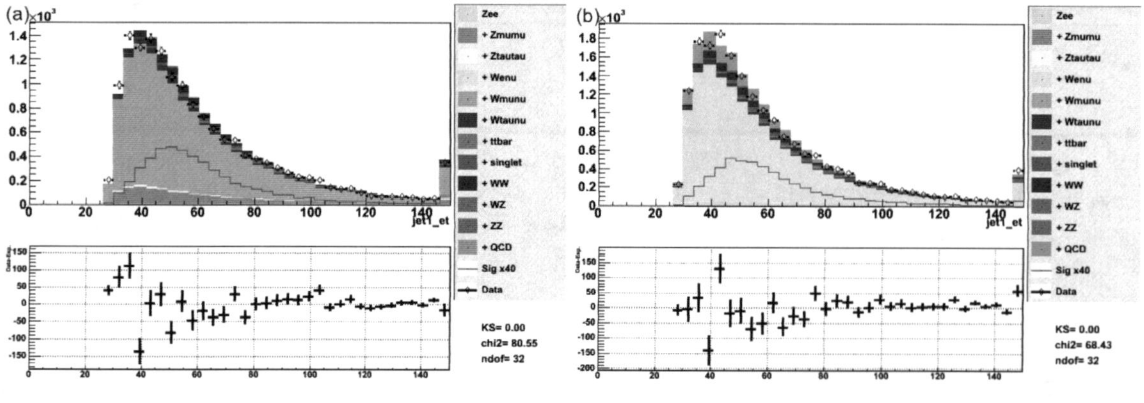

Figure 2. Transverse energy of the leading E_T-jet when the triggered lepton is a muon (a) or a central electron (b).

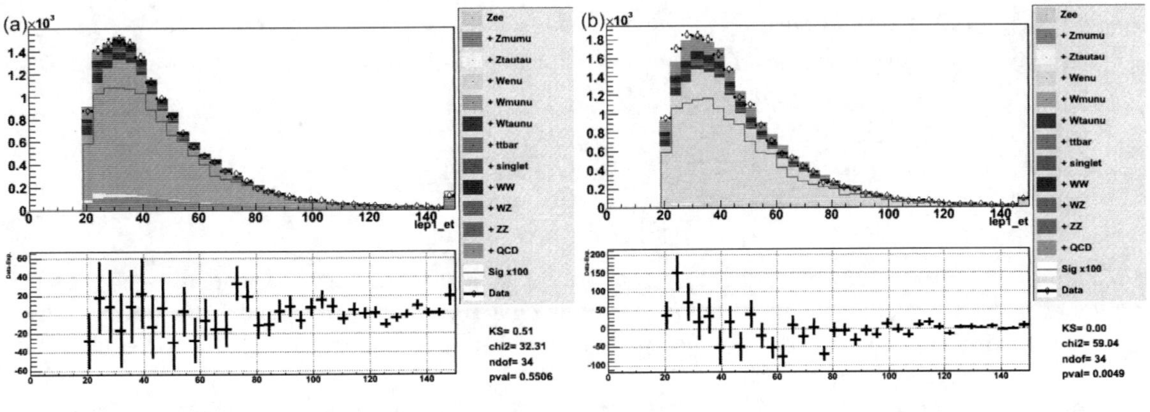

Figure 3. Transverse energy for muons (a) and electrons (b).

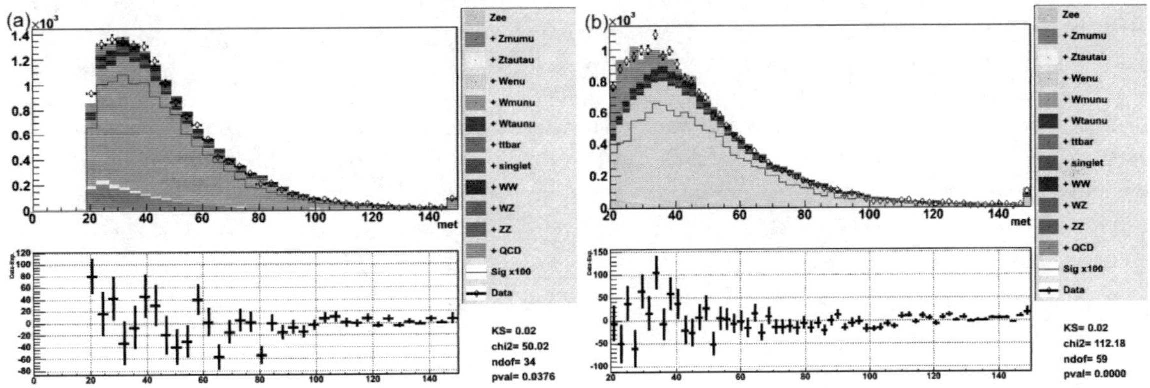

Figure 4. Transverse missing energy when the triggered lepton is a muon (a) or a central electron (b).

3. Jet studies on the WZ→ $\ell\nu 3j$ sample

We would like to select a signal sample among these candidates, and at the same time preserve a fair Z mass resolution, by making use of the information carried by all the jets in the event. In previous studies of the WZ process at CDF the Z mass was defined as $M_{j_1 j_2}$, $j1$, $j2$ being the two leading-E_T jets. In the sample where 3 jets are found $M_{j_1 j_2}$ has a degraded resolution (Fig.5).

3.1. Sample selection

Besides the selection reported in Sec. 2.1 we allow for a third jet with $E_T > 15$ GeV and $|\eta| < 2$.

3.2. Adopted strategy

Improving the resolution in such a sample means to choose the correct jet combination coming from Z for building the Z mass. Our goal is to investigate the origin of the additional jet, which may be initiated by gluon(s) radiated from the interacting partons (ISR) or from the Z-decay products (FSR). In order to do that:

- Jets are ordered in decreasing E_T: j1, j2, j3
- Jets are matched in direction to quarks from Z decay ("MJ")

- When a number of matches different from 2 is found the event is not considered
- Investigate at generator level the origin of the not-matched jet ("NMJ") in order to determine the right jet combination (**"RJC"**)
  (i) NMJ = j3 is from ISR $\mapsto$ RJC = j1j2 - 40% of events
  (ii) NMJ = j2 is from ISR $\mapsto$ RJC = j1j3 - 18% of events
  (iii) NMJ = j1 is from ISR $\mapsto$ RJC = j2j3 - 9% of events
  (iv) NMJ is from FSR $\mapsto$ RJC = j1j2j3 - 19% of events

In 14% of cases two or more jets are not matched. We are planning to rescue these events by implementing a more efficient matching algorithm. This algorithm will search for hadrons rather than quarks in the jet cone and will trace back the origin of the hadrons in order to understand if they were produced by a Z-decay.
Four different Neural Networks (NNs) are trained in MC signal events to isolate each of the above cases. These NNs use the information carried by the kinematical variables and should possibly make us able to decide event by event which is the proper jet combination to be used for building the Z-mass.

3.3. The importance of knowing the correct jet pair
In Fig. 5 we compare M_{j1j2} distribution with the distribution built by using the event-by-event RJC (Sec. 3.2). These distributions have been built in the sample where two jet-to-quark matches are found.

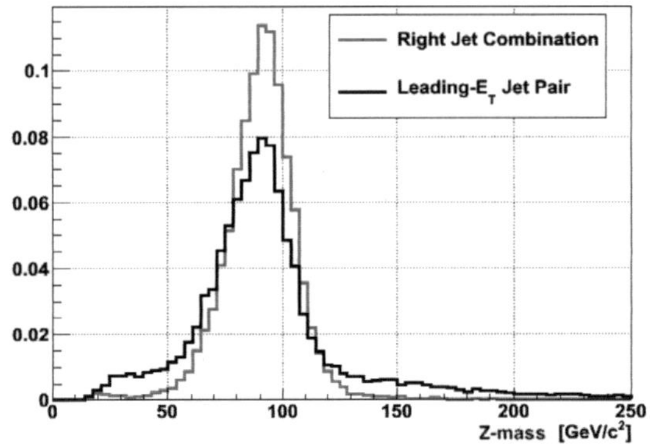

Figure 5. In black, dijet mass built with the two leading jets, while the distribution in blue is built using the RJC.

The low and high mass tails affecting the M_{j1j2} distribution are drastically reduced by choosing the correct combination.

3.4. Look for M_{j1j2}
In this paper only the NN whose goal is to isolate RJC $= j1j2$ from the rest (NN_{12}) will be shown. Results on the other NNs are still preliminary. In order to isolate events when $RJC = j1j2$ we investigate some kinematical variables. We divided the sample into two parts:

- RJC= $j1j2$
- Other jet combinations (RJC = j1j3, j2j3, etc.) which we name "OJC"

Below is the list of the variables used:

(i) $m_{jj'}/m_{j_1j_2j_3}$[2]
(ii) $\gamma_{jj'} = (E_j + E_{j'})/m_{jj'}$
(iii) $\mathrm{d}\eta_{jj'} = |\eta_j - \eta_{j'}|$
(iv) $dR_{jj'} = \sqrt{d\eta^2_{jj'} + d\phi^2_{jj'}}$
(v) $dR_{j_1j_2,j_3}$, dR between the third jet and vectorial sum of two leading jets.
(vi) $dR_{j_1j_2j_3,j_3}$, dR between the third jet and vectorial sum of all the three jets.
(vii) llr, is the likelihood ratio used for discriminating a quark from a gluon jet. For further details see [8].

A subset of the aforementioned observables are shown in Fig. 6. Some additional variables defining jet η and p_T distributions are being studied:

- $p_T(\mathrm{j}_1\text{+}\mathrm{j}_2)/p_T(\mathrm{j}_3)$
- $\eta(j_1\text{+}\mathrm{j}_2)/\eta(\mathrm{j}_3)$

which would distinguish between quark jets and gluon jets.

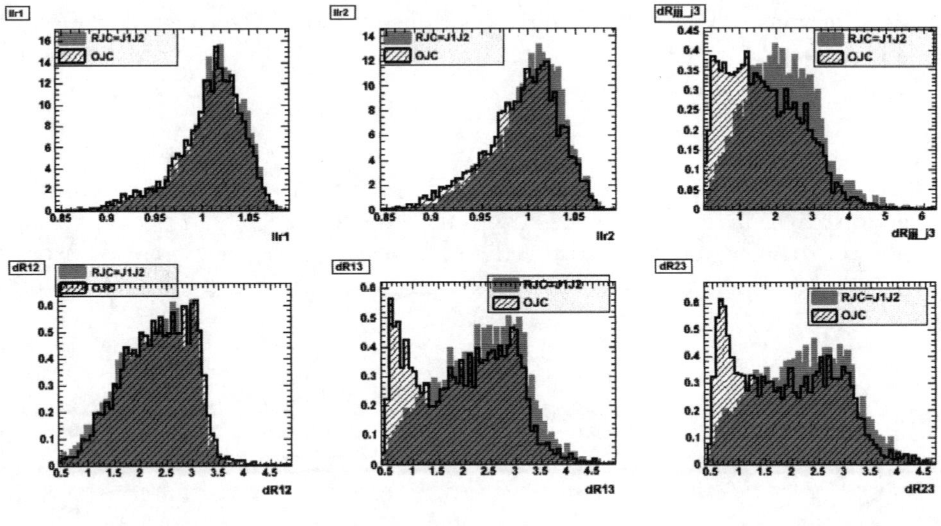

Figure 6. Angular distribution between jet pairs and llr jet distributions. $RJC = j1j2$ is compared to OJC. See text for symbol significance

[2] jj' refers to the three possible combinations: j1j2, j1j3 and j2j3.

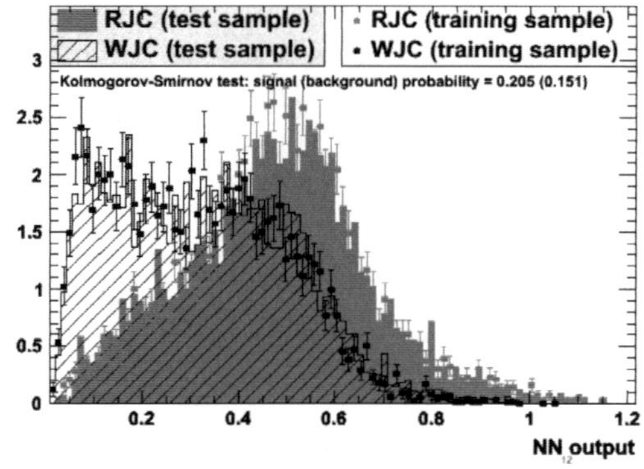

Figure 7. NN_{12} output

3.5. NN_{12} output

In order to avoid background (W+jets, $t\bar{t}$, etc...) being sculpted later, the input variables are weighted. Weights applied are calculated such that the M_{j1j2} distribution in the OJC sample become approximately the same of the one in the $RJC = j1j2$ sample. By doing this we will decorrelate the NN_{12} output from the numerical value of M_{j1j2} and make it sensitive only to the kinematical distributions of the involved variables. The variables described above are weighted accordingly and are used for training a Neural Network. We employ the MLP method, (see [9] for details). The NN_{12} response is shown in Fig. 7.

3.6. Results

In order to understand the impact of this method on the sensitivity of the measurement we apply the NN_{12} selection on the major sources of background (W+jets, Z+jets, $t\bar{t}$ and single top) events and compare them to WZ events. In Fig. 8 the standard and the new mass distributions are shown in both signal and background events. The signal is multiplied by 70 in order to facilitate a visual comparison.

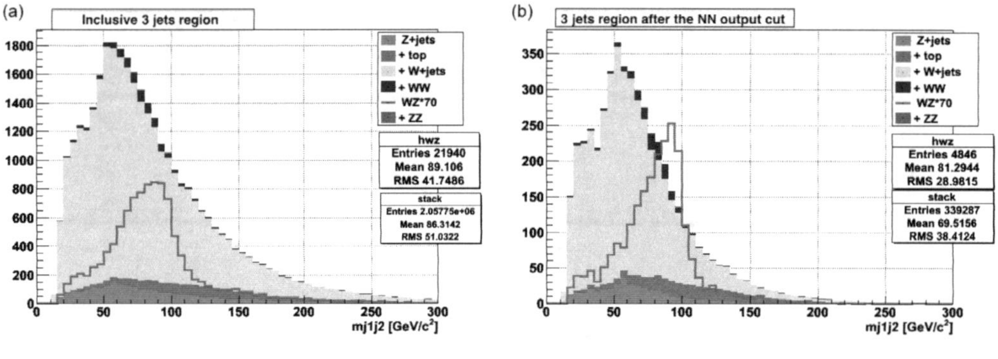

Figure 8. (a) Mj1j2. (b) Mj1j2 if we require $NN_{12} \geq 0.5$

We note the better separation when a cut on the NN_{12} output is applied. We note that as a consequence of the above mentioned decorrelation procedure, the M_{j1j2} distribution in the background events doesn't get sculpted. Therefore the S/B ratio increases, where S, B are calculated in the selected mass window [70,110] GeV/c^2.

Table 2. S, B are calculated in the selected mass window [70,110] GeV/c^2. "G" gives the σ/μ ratio in the mass window, where σ and μ are the width and the mean of the gaussian distribution used for the fit.

	std	if NN_{12} >0.5
S	79.59	6.52
B	10322	329
S/ B	0.008	0.02
G for Evt/Wind	16.22 %	12.3%

3.7. Conclusions

After the NN_{12} output cut, we select 25% of WZ events in which Z peak is appreciably narrower. The resolution is improved by 25% and signal over background is improved by a factor of 2.
We plan to repeat the procedure to isolate events where RJC = j1j3; j2j3; j1j2j3 and use the information carried by these four NNs for deciding event by event which is the proper combination for building the Z mass.
If successfull, the technique will significantly increase the signal acceptance of several important analyses (Diboson, Higgs, etc.). We note that when applied to the b-tagged sample for the light Higgs search, the performance of this method is expected to change because of the different background composition (mostly $t\bar{t}$) and of the different fragmentation properties of b-flavored jets.

Acknowledgments

We are grateful to Prof. Giorgio Bellettini, Dr. Giuseppe Latino, Dr. Vadim Rusu and Dr. George Velev for many fruitful discussions and suggestions.

References

[1] Hagiwara K et al 1990 *Phys. Rev.* D **41** 2113
[2] Abazov V M et al (D0 collaboration), arXiv:0904.0673v1.
[3] Culbertson R et al *Search for Anomalous Production of di-photon+MET Events in 2/fb of Data* CDF **9184**
[4] Junk T et al *Combined CDF and D0 Upper Limits on Standard Model Higgs Boson Production with up to 6.7 fb^{-1}* CDF **10241**
[5] Mangano M L, Moretti M, Piccinini F, Pittau R, and Polosa A 2001 *J. High Energy Phys.* **07** 001.
[6] Sjöstrand T, Edén P, Friberg C, Lönnblad L, Miu G, Mrenna S and Norrbin E 2001 *Computer Phys. Commun.* **135** 238.
[7] B. Cooper, A. Messina *Estimation of the Background to W→ eν+n jets events* CDF **6636**.
[8] Ketchum W, Rusu V, Alawi Z, Young-Kee K *Search for ZW/ZZ Production in Leptons + Jets channel* CDF **10061**.
[9] Hoecker A, Speckmayer P, Stelzer J, Therhaag J, von Toerne E, Voss H 2009 *TMVAUsersGuide* http: tmva.sourceforge.net

Vector boson productions and Higgs searches in pp collisions at $\sqrt{s} = 7$ TeV

Jónatan Piedra for the CMS Collaboration

E-mail: piedra@cern.ch

Abstract. Measurements of Wγ, Zγ and WW production in proton-proton collisions at $\sqrt{s} = 7$ TeV are presented, together with searches for the Higgs boson. The results are based on a data sample corresponding to an integrated luminosity of 36 pb^{-1} recorded by the CMS experiment. In the Wγ, Zγ and WW cross sections measurements the electron and muon decay channels of the W and Z are used. A search for the Higgs boson is performed in the fully leptonic WW final state, not revealing any evidence of excess above backgrounds. Limits are set on the production of the Higgs boson in the context of the SM and in the presence of a sequential fourth family of fermions with high masses. A MSSM Higgs boson search is perforfmed in the tau pairs decay. No excess is observed in the tau-pair invariant mass spectrum.

1. The CMS detector [1]

The central part of the CMS detector is a superconducting solenoid, which provides an axial magnetic field of 3.8 T parallel to the beam axis. Charged particle trajectories are measured by the silicon pixel and strip tracker, which covers the pseudorapidity region $|\eta| < 2.5$. A crystal electromagnetic calorimeter and a brass/scintillator hadron calorimeter surround the tracking volume and cover $|\eta| < 3.0$. A quartz-fiber Cherenkov calorimeter extends the coverage to $|\eta| = 5.0$. Muons are measured in gas detectors embedded in the iron return yoke outside the solenoid, in the pseudorapidity range $|\eta| < 2.4$. The detector is nearly hermetic, allowing for energy balance measurements in the plane transverse to the beam axis.

2. Measurement of Wγ and Zγ production [2]

The study of Wγ and Zγ production in proton-proton collisions is an important test of the Standard Model (SM) because of its sensitivity to the self-interaction between gauge bosons via trilinear gauge boson couplings (TGCs). Any deviation of the observed strength of the TGC from the SM prediction would indicate new physics. The Wγ and Zγ processes are studied in the final states $\ell\nu\gamma$ and $\ell\ell\gamma$, respectively, where ℓ is either an electron or a muon. Leading order (LO) Wγ production can be described by three processes: initial state radiation (ISR), where a photon is radiated by one of the incoming quarks; final state radiation (FSR), where a photon is radiated from the charged lepton from the W boson decay; and finally through the WWγ vertex, where a photon couples directly to the W boson. In the SM, LO Zγ production is described via ISR and FSR processes only, because the ZZγ and Z$\gamma\gamma$ TGCs are not allowed at the tree level. As at LO the Wγ and Zγ cross sections diverge for soft photons or, in the case of Z/$\gamma^*\gamma$ production, for small values of the dilepton invariant mass, we restrict the cross section measurement to the phase space defined by the following two kinematic requirements:

the photon candidate must have transverse energy E_T^γ larger than 10 GeV, and it must be spatially separated from the final-state charged lepton(s) by $\Delta R(\ell, \gamma) > 0.7$. Furthermore, for the Z$\gamma$ final state, the invariant mass of the two lepton candidates must be above 50 GeV. The main background to Wγ and Zγ production consists of W+jets and Z+jets events, respectively, where the photon candidate originates from one of the jets. We estimate this background from data. The contribution from other processes, such as $\mathrm{t\bar{t}}$ and multijet QCD production, is much smaller and it is estimated from Monte Carlo (MC) simulation studies.

The details of the photon, electron and muon reconstruction, together with the calculation of the transverse missing energy E_T^{miss}, can be found in [2]. To estimate the background due to jets misidentified as photons we measure the E_T^γ-dependent ratio of jets passing the full photon identification criteria to those identified as photons but failing the track isolation requirement. The obtained probability is folded with the non-isolated photon candidates in the Wγ and Zγ candidate events to estimate the number of W+jets and Z+jets events, respectively, passing the full selection criteria. This estimation is further cross-checked with W+jets and Z+jets simulation and with the results obtained from an independent study of photon cluster shower shapes [2]. We observe good agreement between all three methods (Fig. 1).

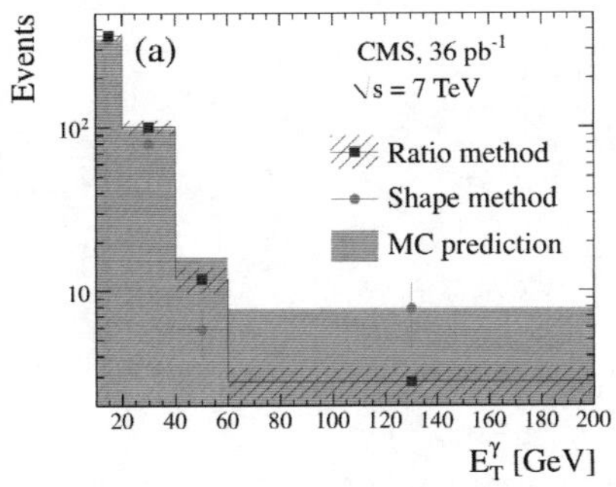

Figure 1. The background from misidentified jets as a function of the photon candidate transverse energy estimated from the ratio method is shown with blue squares, together with an alternative method that uses energy deposition shape templates (magenta circles), and simulation (green filled histogram) for the Wγ channel. Uncertainties include both statistical and systematic sources.

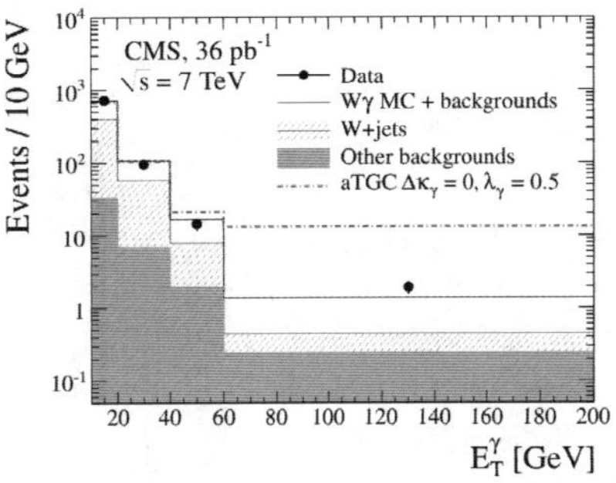

Figure 2. Transverse energy distribution for the photon candidates for Wγ production. Data are shown with black circles; expected signal plus background is shown as a black solid line; the contribution from misidentified jets is given as a hatched blue histogram, and the background from γ+jets, $\mathrm{t\bar{t}}$, and multiboson processes is given as a solid green histogram. A typical anomalous TGC signal is given as a dot-dashed red line.

The W$\gamma \rightarrow \ell\nu\gamma$ final state is characterized by a prompt, energetic and isolated lepton, significant E_T^{miss} due to the presence of the neutrino and a prompt isolated photon. The event selection is similar for the electron and muon channels: we require a charged lepton with $p_T > 20$ GeV which must satisfy the trigger requirements; one photon with transverse energy $E_T^\gamma > 10$ GeV, and the E_T^{miss} in the event exceeding 25 GeV. After the full selection, 452 events are selected in the e$\nu\gamma$ channel and 520 events are selected in the $\mu\nu\gamma$ channel. No events have more than one photon candidate in the final state. The background from misidentified jets estimated in data amounts to $220 \pm 16\,(\mathrm{stat}) \pm 14\,(\mathrm{syst})$ and $261 \pm 19\,(\mathrm{stat}) \pm 16\,(\mathrm{syst})$ events, respectively. The E_T distribution for photon candidates in events passing the full Wγ selection is given in Fig. 2. Events in the Zγ sample are selected by requiring a pair of electrons or muons, each with transverse momentum $p_T > 20$ GeV, forming an invariant mass above 50 GeV. One of these leptons must satisfy the trigger requirements. The events are further required to have a photon candidate passing the selection criteria with transverse energy above 10 GeV. We observe 81 events in the eeγ final state and 90 events in the $\mu\mu\gamma$ final state. No events are observed with more than one photon candidate. The Z+jets background to these final states is estimated to be $20.5 \pm 1.7\,(\text{stat}) \pm 1.9\,(\text{syst})$ and $27.3 \pm 2.2\,(\text{stat}) \pm 2.3\,(\text{syst})$, respectively. The distribution of the $\ell\ell\gamma$ mass as a function of the dilepton mass is displayed in Fig. 3.

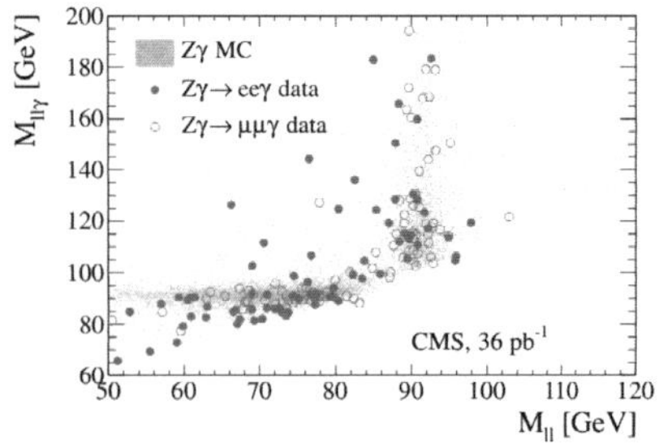

Figure 3. Distribution of the $\ell\ell\gamma$ invariant mass as a function of the dilepton invariant mass for selected Zγ candidates in the electron (filled circles) and muon (open circles) final states. The data accumulation at $M_{\ell\ell\gamma} \simeq M_Z$ corresponds to FSR events, while the data at $M_{\ell\ell} \simeq M_Z$ correspond to ISR events.

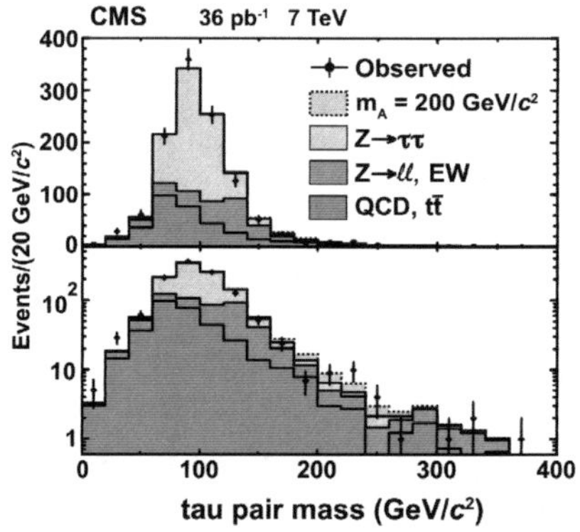

Figure 4. The reconstructed tau-pair invariant mass on linear (above) and logarithmic (below) scales, for the sum of the eτ_h, $\mu\tau_h$ and eμ final states, comparing the observed data points to the expected backgrounds. The contribution from a Higgs signal is also shown.

The systematic uncertainties considered include uncertainties on lepton and photon energy scales and resolution, effects from pile-up interactions, and uncertainties in the parton distribution functions; uncertainties affecting the data vs. simulation correction factors for the efficiencies of the lepton trigger, lepton and photon reconstruction and identification, and E_T^{miss} efficiencies for the Wγ process; uncertainties on the data-driven W+jets and Z+jets background estimation; and finally, an additional uncertainty due to the measurement of the integrated luminosity. For $E_T^{\gamma} > 10$ GeV, $\Delta R(\ell,\gamma) > 0.7$), and $m_{\ell\ell} > 50$ GeV for the Zγ process, we find the cross sections for Wγ and Zγ production to be in good agreement with the next-to-leading order (NLO) predictions of 49.4 ± 3.8 pb and 9.6 ± 0.4 pb, respectively:

$$\begin{aligned}
\sigma(\text{pp} \to \text{W}\gamma + \text{X}) \times \mathcal{B}(\text{W} \to \text{e}\nu) &= 57.1 \pm 6.9\,(\text{stat}) \pm 5.1\,(\text{syst}) \pm 2.3\,(\text{lumi})\text{ pb},\\
\sigma(\text{pp} \to \text{W}\gamma + \text{X}) \times \mathcal{B}(\text{W} \to \mu\nu) &= 55.4 \pm 7.2\,(\text{stat}) \pm 5.0\,(\text{syst}) \pm 2.2\,(\text{lumi})\text{ pb},\\
\sigma(\text{pp} \to \text{Z}\gamma + \text{X}) \times \mathcal{B}(\text{Z} \to \text{ee}) &= 9.5 \pm 1.4\,(\text{stat}) \pm 0.7\,(\text{syst}) \pm 0.4\,(\text{lumi})\text{ pb},\\
\sigma(\text{pp} \to \text{Z}\gamma + \text{X}) \times \mathcal{B}(\text{Z} \to \mu\mu) &= 9.2 \pm 1.4\,(\text{stat}) \pm 0.6\,(\text{syst}) \pm 0.4\,(\text{lumi})\text{ pb}.
\end{aligned}$$

3. Measurement of W^+W^- production and search for the Higgs boson [4]

The origin of the masses of W and Z bosons is attributed, in the SM, to the spontaneous breaking of electroweak symmetry caused by a new scalar field. The existence of the associated field quantum, the Higgs boson, has yet to be experimentally confirmed. A possible extension to the SM is the addition of a fourth family of fermions. For sufficiently large lepton and quark masses, this extension has not been excluded by existing constraints. The presence of another fermion family produces an enhancement of the dominant gluon fusion cross section, together with some changes in the Higgs decay branching fractions. The choice of infinitely heavy quarks of the fourth family in the extended SM yields to the smallest enhancement factor for the Higgs boson cross section, hence to the most conservative scenario for the exclusion of such a model. This scenario is used to set limits in this paper.

The dominant irreducible background for $H \to WW$ production is the SM non-resonant production of WW. A good understanding of this process and its properties is thus needed for the Higgs boson search. The WW production has been extensively studied by the LEP and Tevatron experiments, where it has been found to be in agreement with the SM prediction. In pp collisions at the LHC, the SM NLO QCD prediction of the WW production cross section at $\sqrt{s} = 7$ TeV is 43.0 ± 2.0 pb. The WW candidates, with both W bosons decaying leptonically, are selected in final states consisting of two isolated, high transverse momentum, oppositely-charged leptons (electrons or muons), and large missing transverse energy due to the undetected neutrinos. Events are selected in three final states: ee, $\mu\mu$ and eμ, which include $W \to \tau\nu_\tau$ events with leptonic τ decays. The details of the lepton and jet reconstruction, and missing transverse energy calculation can be found in [4]. For the event selection also a derived quantity called *projected* E_T^{miss} is used. With $\Delta\phi$ the azimuthal angle between the E_T^{miss} and the closest lepton, the *projected* E_T^{miss} is defined as the component of E_T^{miss} transverse to the lepton direction if $\Delta\phi$ is smaller than $\pi/2$, and the full magnitude of E_T^{miss} otherwise. This variable, which helps to reject $Z \to \tau\tau$ background events, is required to be above 35 GeV in the ee and $\mu\mu$ final states, and above 20 GeV in the eμ final state. To further reduce the Drell–Yan background in the ee and $\mu\mu$ final states, events with a dilepton invariant mass within 15 GeV of the Z mass are discarded. Events are also rejected with dilepton masses below 12 GeV, to suppress contributions from low mass resonances.

To reduce backgrounds containing top quarks, events containing at least one jet with $|\eta| < 5.0$ and $p_T > 25$ GeV are rejected. In addition, this jet veto is complemented by a top veto based on soft-muon and b-jet tagging. To reduce the background from diboson processes, such as WZ and ZZ production, any event which has an additional third lepton passing the identification and isolation requirements is rejected. Figure 5 shows the jet multiplicity distribution for events that pass all selections but the jet veto and top veto. Figure 6 shows the dilepton mass distribution for events passing the final WW event selections, except the Z mass veto. The W+jets background is estimated using data-driven methods by defining a set of loosely selected lepton-like objects in a sample of events dominated by dijet production. The total misidentified electron and muon background contributions are found to be $1.2 \pm 0.3\,(\text{stat}) \pm 0.6\,(\text{syst})$ and $0.5 \pm 0.3\,(\text{stat}) \pm 0.3\,(\text{syst})$ events, respectively. The remaining top quark background after the full event selection can be estimated from data by counting events with either an additional soft muon or at least one b-tagged jet with p_T below the jet veto threshold. No events are rejected by the top veto in data after applying the full selection, which is consistent with the prediction from simulation. Therefore, the top quark background contribution is taken directly from simulation, which predicts $0.77 \pm 0.05\,(\text{stat}) \pm 0.77\,(\text{syst})$ events, where a 100% systematic uncertainty is assigned as a conservative estimate of the difference between data and simulation. An estimate of the residual Z boson contribution in the ee and $\mu\mu$ final states outside the Z mass window is obtained from data in the following way: the ratio of the number of events outside the Z mass window to that inside is obtained from simulation. The observed number of events inside the

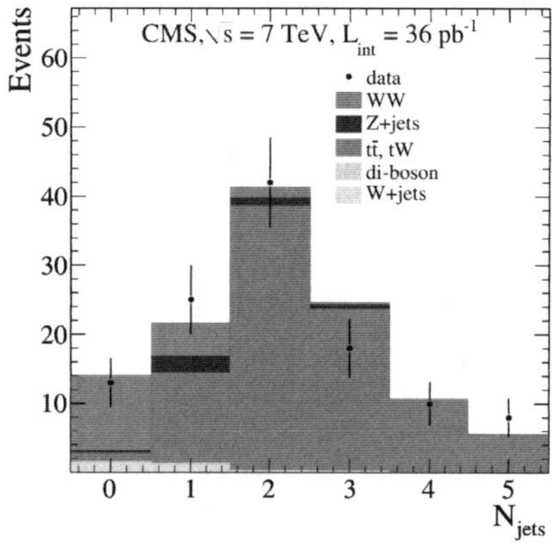

Figure 5. Jet multiplicity distribution after all WW selection criteria, except the top veto and jet veto requirements.

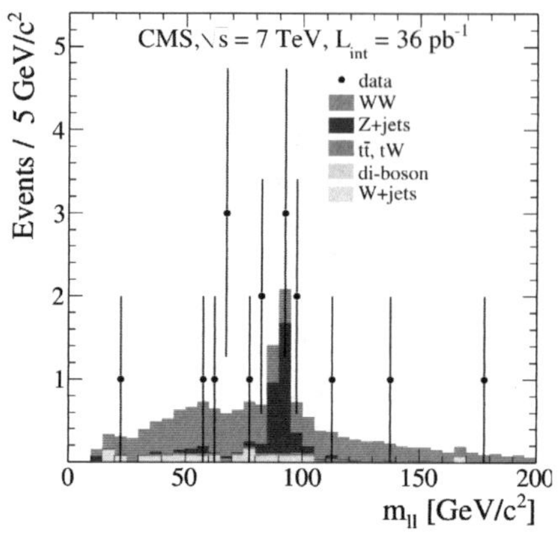

Figure 6. Dilepton mass distribution for the events passing the final WW selection, except the Z mass veto.

Z mass window in data, from which the non-Z contributions are subtracted, is then scaled by this ratio to compute the residual Z background, estimated as $0.2 \pm 0.2\,(\text{stat}) \pm 0.3\,(\text{syst})$ events. Other backgrounds are estimated from simulation.

The WW signal efficiency is estimated using the simulation, corrected by data-to-simulation scale factors. For electron and muon reconstruction and identification, a tag-and-probe is applied to leptons, both in data and simulation. For estimating the effect of the jet veto efficiency on the WW signal, events in the Z resonance region are used. Overall, the uncertainty is estimated to be 7% on the WW selection efficiency, coming mainly from the theoretical uncertainty in the jet veto efficiency determination. The uncertainty on the background estimations in the WW signal region is about 37%. The WW yield is calculated from the number of events in the signal region, after subtracting the expected contributions of the various SM background processes. From this yield the WW cross section is found to be

$$\sigma(\text{pp} \to \text{WW} + \text{X}) \times \mathcal{B}(\text{W} \to \ell\nu) \quad = \quad 41.1 \pm 15.3\,(\text{stat}) \pm 5.8\,(\text{syst}) \pm 4.5\,(\text{lumi})\ \text{pb}.$$

The preselection for the Higgs boson search in the WW decay mode is identical to the WW described selection. To enhance the sensitivity to the Higgs boson signal, two different analyses are performed. The first analysis is a cut-based approach where further requirements on a few observables are applied, while the second analysis makes use of multivariate techniques. In the cut-based approach, the extra selections are based on the transverse momenta of the two leptons, the dilepton invariant mass and the azimuthal angular separation between the two leptons, $\Delta\phi_{\ell\ell}$. Figure 7 shows the $\Delta\phi_{\ell\ell}$ distribution after applying the WW selection, for a SM Higgs boson signal with $m_{\text{H}} = 160$ GeV. In the multivariate approach a boosted decision tree (BDT) technique is used for each Higgs boson mass hypothesis. The non-resonant WW contribution in the $\text{H} \to \text{WW}$ signal region is estimated from data using the dilepton mass distribution. For a given Higgs boson mass, the region with a small contribution from Higgs boson decays is selected and simulation is used to extrapolate the WW background into the signal region. In addition, for the present measurement other Higgs boson production mechanisms are considered as backgrounds: a Higgs boson in the final state accompanied by a W or Z boson or by a pair of top quarks, and the vector boson fusion process. These processes are heavily suppressed by the jet and additional lepton veto requirements, and the corresponding yield

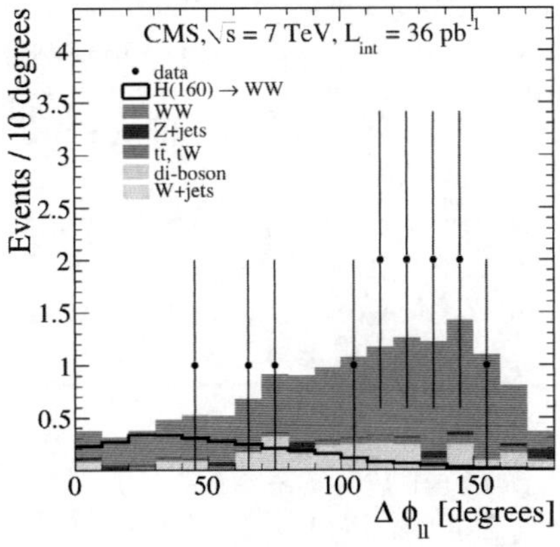

Figure 7. Azimuthal angular separation between the two leptons after the WW selection, for $m_{\mathrm{H}} = 160$ GeV. The area marked as WW corresponds to the non-resonant contribution.

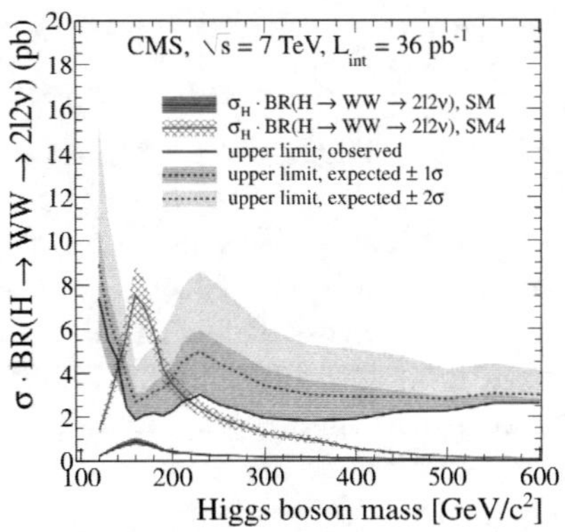

Figure 8. 95% mean expected and observed C.L. upper limits on $\sigma_{\mathrm{H}} \times \mathcal{B}$ using multivariate BDT event selections. Results are obtained using a Bayesian approach.

amounts to 1–2% of the gluon fusion process. Systematic uncertainties related to acceptance and efficiencies for $\mathrm{H} \rightarrow \mathrm{WW}$ are estimated in a similar way as described in the WW cross section measurement. The overall uncertainty on the $\mathrm{H} \rightarrow \mathrm{WW}$ signal yield is estimated to be of about 14%, where the uncertainty on the jet veto efficiency and the luminosity determination are the main contributions. The uncertainties on the background estimations are about 40%.

Upper limits are derived on the product of the gluon fusion Higgs boson production cross section by the $\mathrm{H} \rightarrow \mathrm{WW}$ branching fraction. Two different statistical methods are used, both using the same likelihood function from the expected number of observed events modeled as a Poisson random variable whose mean value is the sum of the contributions from signal and background processes. The first method is based on Bayesian inference, while the second method, known as CL_s, is based on the hybrid Frequentist-Bayesian approach. Both methods account for systematic uncertainties. The upper limits for the Bayesian inference are about 1-3% higher than the CL_s method. Results are reported in the following using only the Bayesian approach, with a flat signal *prior*. The 95% observed and mean expected C.L. upper limits are shown in Fig. 8 for Higgs boson masses in the range 120 – 600 GeV. The $\sigma_{\mathrm{H}} \times \mathcal{B}(\mathrm{H} \rightarrow \mathrm{WW} \rightarrow 2\ell 2\nu)$ upper limits are about three times larger than the SM expectation for $m_{\mathrm{H}} = 160$ GeV. When compared with recent theoretical calculations performed in the context of a SM extension by a sequential fourth family of fermions with very high masses, the results of BDT analyses exclude at 95% C.L. a Higgs boson with mass in the range from 144 to 207 GeV, where the mean expected exclusion at 95% C.L. is in the range from 147 to 193 GeV. Similar results are achieved using the cut-based approach.

4. Search for neutral MSSM Higgs bosons decaying to tau pairs [5]

The SM Higgs boson suffers from quadratically divergent self-energy corrections at high energies. Numerous extensions to the SM have been proposed to address these divergences. One such model, supersymmetry, a symmetry between fundamental bosons and fermions, results in cancellation of the divergences at tree level. The minimal supersymmetric extension to the standard model (MSSM) requires the presence of two Higgs doublets. This leads to a more complicated scalar sector, with five massive Higgs bosons: a light neutral state (h), two charged states ($H^{\pm}$), a heavy neutral CP-even state (H) and a neutral CP-odd state (A). The masses of the MSSM Higgs boson states are specified up to radiative corrections mainly by two parameters, usually taken to be the mass of the pseudoscalar state, m_A, and the ratio of the vacuum expectation values of the two Higgs doublets, $\tan\beta$. At large $\tan\beta$ (greater than about 20–30), the couplings of the Higgs bosons to down-type quarks are approximately proportional to $\tan\beta$. As a result, the production cross section for two of the three neutral Higgs bosons can be nearly as large as that for the electroweak gauge bosons W and Z at a pp collider such as the LHC. Two main production processes contribute to $pp \to \phi + X$, where $\phi =$ h, H or A: gluon fusion through a b-quark loop and direct $b\bar{b}$ annihilation from the b-parton density in the beam protons. The tau-pair decays of the neutral Higgs bosons, having a branching fraction of roughly 10%, serve as the best experimental signature for this search. The $b\bar{b}$ mode, though it has a much larger branching fraction, suffers from an overwhelming background from QCD processes. Three final states where one or both taus decay leptonically are used: $e\tau_h$, $\mu\tau_h$, and $e\mu$, where we use the symbol τ_h to indicate a reconstructed hadronic decay of a τ.

The main challenge in the identification of hadronic tau decays is overcoming the large background due to hadronic jets from QCD processes. Hadronic tau decays almost always yield one or three charged pions, plus zero to several neutral pions, depending on the decay mode. The algorithm used here starts with a high-p_T reconstructed charged hadron, and combines it with other nearby reconstructed charged hadron and neutral pion candidates. The algorithm considers all possible combinations of these objects and determines which are consistent with the kinematics of tau decay. Among those, it chooses the most isolated in terms of the presence of nearby reconstructed particles. For the $\mu\tau_h$ and $e\tau_h$ final states, we select events with an isolated muon or electron with $p_T > 15$ GeV and $|\eta| < 2.1$, and an oppositely charged τ_h with $p_T > 20$ GeV and $|\eta| < 2.3$. The transverse mass M_T of the $\ell = e, \mu$ with the missing transverse energy of the event is required to be below 40 GeV, in order to reduce the background from W+jets events. For the $e\mu$ final state, we select events with an isolated electron with $|\eta| < 2.5$ and an oppositely charged isolated muon with $|\eta| < 2.1$, both with $p_T > 15$ GeV and $M_T <$ 50 GeV (to reject WW and $t\bar{t}$ events), calculated for each lepton separately. We reject events in which there are more than one e or μ.

The largest source of events selected with these requirements comes from $Z \to \tau\tau$. We estimate this contribution with simulation, and we determine its normalization based on the number of observed $Z \to ee$ and $Z \to \mu\mu$ events. A significant source of background arises from QCD multijet events and W+jets events in which a jet is misidentified as τ_h, and there is a real or misidentified e or μ. The rates for these processes are estimated using the number of observed same-charge events, and cross-checked using the jet-to-tau misidentification rate measured in multijet events. Other background processes include $t\bar{t}$ production and $Z \to ee/\mu\mu$ events.

To distinguish the Higgs boson signal from the background, we reconstruct the tau-pair mass using a likelihood technique. The algorithm estimates the original tau three-momenta by maximizing a likelihood with respect to free parameters corresponding to the missing tau-neutrino momenta, and subject to all applicable kinematic constraints. Other terms in the likelihood take into account the tau-decay phase space and the probability density in the tau transverse momentum, parametrized as a function of the tau-pair mass. The observed reconstructed tau-pair mass distribution summed over all three channels is shown in Fig. 4. Various effects can alter the shape and normalization of the reconstructed tau-pair invariant mass spectrum. The main sources include the total integrated luminosity, background normalizations, Z production cross section, and lepton identification and isolation efficiency. Uncertainties that contribute to mass spectrum shape variations include the lepton energy scales, and uncertainties on the E_T^{miss} scale that is used for the tau-pair invariant mass reconstruction. To search for the presence of a Higgs boson signal in the selected events, we perform a maximum likelihood fit to the tau-pair invariant mass spectrum. Systematic uncertainties are represented by nuisance parameters, which we remove by marginalization, assuming a log normal prior for normalization parameters, and Gaussian priors for mass-spectrum shape uncertainties. The mass spectra show no evidence for the presence of a Higgs boson signal, and we set 95% CL upper bounds on the Higgs boson cross section times the tau-pair branching fraction.

5. Conclusions

In a data sample corresponding to an integrated luminosity of 36 pb^{-1} of pp collisions at $\sqrt{s} = 7$ TeV, we have presented the first measurement of the Wγ and Zγ cross sections, for $E_T^{\gamma} > 10$ GeV, $\Delta R(\gamma, \ell) > 0.7$, and for the additional requirement on the dilepton invariant mass to exceed 50 GeV for the Zγ process. The results are in good agreement with the NLO prediction values. We have presented the first measurement of the WW cross section and a search for the Higgs boson decaying to WW. The WW events have been used to measure the WWγ and WWZ triple gauge couplings. The results, which are in agreement with the SM predictions, are consistent with the precise measurements made at LEP and comparable with the current Tevatron results. Limits on the Higgs boson production cross section have been derived. No excess above the SM expectations was found. In the presence of a sequential fourth family of fermions with very high masses, a Higgs boson with SM couplings and a mass between 144 and 207 GeV has been excluded at 95% confidence level, using a Bayesian inference method. Finally, we have performed a search for neutral MSSM Higgs bosons. The tau-pair decay mode in final states with one e or μ plus a hadronic decay of a tau, and the eμ final state were used. The observed tau-pair mass spectrum reveals no evidence for neutral Higgs boson production, and we determine an upper bound on the product of the Higgs boson cross section and tau-pair branching fraction as a function of m_{A}. These results, interpreted in the MSSM parameter space of $\tan\beta$ versus m_{A}, in the $m_{\mathrm{h}^{\mathrm{max}}}$ scenario, exclude a previously unexplored region reaching as low as $\tan\beta = 23$ at $m_{\mathrm{A}} = 130$ GeV.

Acknowledgments

We wish to congratulate our colleagues in the CERN accelerator departments for the excellent performance of the LHC machine. We thank the technical and administrative staff at CERN and other CMS institutes, and acknowledge support from: FMSR (Austria); FNRS and FWO (Belgium); CNPq, CAPES, FAPERJ, and FAPESP (Brazil); MES (Bulgaria); CERN; CAS, MoST, and NSFC (China); COLCIENCIAS (Colombia); MSES (Croatia); RPF (Cyprus); Academy of Sciences and NICPB (Estonia); Academy of Finland, ME, and HIP (Finland); CEA and CNRS/IN2P3 (France); BMBF, DFG, and HGF (Germany); GSRT (Greece); OTKA and NKTH (Hungary); DAE and DST (India); IPM (Iran); SFI (Ireland); INFN (Italy); NRF and WCU (Korea); LAS (Lithuania); CINVESTAV, CONACYT, SEP, and UASLP-FAI (Mexico); PAEC (Pakistan); SCSR (Poland); FCT (Portugal); JINR (Armenia, Belarus, Georgia, Ukraine, Uzbekistan); MST and MAE (Russia); MSTD (Serbia); MICINN and CPAN (Spain); Swiss Funding Agencies (Switzerland); NSC (Taipei); TUBITAK and TAEK (Turkey); STFC (United Kingdom); DOE and NSF (USA).

References

[1] CMS Collaboration doi:10.1088/1748-0221/3/08/S08004 JINST **3** (2008) S08004
[2] CMS Collaboration CERN-PH-EP-2011-045 arXiv:1105.2758
[3] CMS Collaboration CMS-DP-2011-002 http://cdsweb.cern.ch/record/1335668
[4] CMS Collaboration CERN-PH-EP-2011-015 arXiv:1102.5429 Phys. Lett. **B** 699 (2011) 25-47
[5] CMS Collaboration CERN-PH-EP-2011-027 arXiv:1104.1619

Experiences on QCD Monte Carlo simulations: a user point of view on the inclusive jet cross-section simulations

Paolo Francavilla

INFN Sez. di Pisa, University of Pisa, School of Graduate Studies G. Galilei
CERN

E-mail: paolo.francavilla@cern.ch

Abstract. In the last years, important progresses in the theoretical description of the QCD high p_T processes have been carried out. In this proceeding, a review of the tools and techniques used to simulate QCD cross sections will be presented from a user point of view. The benchmark process selected for the discussion is the inclusive jet cross section. The proceeding will focus on the uncertainties of the Next to Leading Order (NLO) cross sections, on the strategies adopted to correct for the non-perturbative effects such as the hadronization and the underlying event, and the new techniques derived during the last years to incorporate in a coherent way the NLO matrix elements in the Monte Carlo generators.

1. Introduction

The imminent advent of the measurements at the Large Hadron Collider (LHC) guided important progresses to automatize the theoretical prediction of the Standard Model processes in hadron colliders in the last years. Some of them are oriented at the development of efficient methods to estimate the Next to Leading Order (NLO) - or in some cases even the higher orders - contributions for a big number of different processes. Other important progresses are aimed at merging the exact perturbative expansion approach to approximate models, trying to incorporate in the final prediction part of the contributions of all the higher orders. This proceeding is not a complete review of these techniques, but a report of the hands-on experiences in estimating the theoretical prediction for a specific process.

The process under investigation is the jet production at Large Hadron Collider (LHC), in proton-proton collisions. The jets are a spray of collimated particles produced in high energy collisions. Their production is the dominant high transverse-momentum (p_T) process at the hadron colliders, and it gives us the first glimpse of physics at the TeV scale.

Furthermore, the LHC collaborations have recently reported the measurement of the inclusive jet cross section with the 2010 data [1, 2, 3]. This gives the possibility to directly test the predictions against the measured spectrum.

2. Fixed order predictions

Fixed-order predictions, which involve the first terms in the Quantum Chromo-Dynamics (QCD) perturbative expansion for a given observable, are conceptually quite simple: it is easy to state which contributions are included, and as one includes further orders in the expansion one can reasonably hope to see systematic improvement in the accuracy of ones predictions. The expansion is in powers of the strong coupling constant α_S. Usually, only the first 2 or 3 terms in the expansion are known. The first term is known as the Leading Order (LO).

2.1. Leading Order

The LO theoretical prediction for the inclusive jet cross section is done calculating the QCD 2→2 matrix elements.

In this calculation, the initial protons are replaced by the initial partons (quarks or gluons). The two initial partons carry a fraction of the initial protons momentum, which is statistically described by the proton Parton Density Functions (PDF). The spatial momentum of these incoming partons have only a longitudinal component (parallel to the beam axis), and no transverse momentum. At the LO the proton remnants are not taken into account, with the assumption that their contribution to the hard scattering is minimal.

Once the initial partons are extracted from the proton, the QCD amplitude for the 2→2 parton processes can be calculated. The number of sub-processes which must be taken into account is high, and they can be divided according to the initial partonic configuration (i.e. gluon-gluon, quark-gluon, quark-quark).

The final state of the LO calculation consists of two outgoing partons. They are back-to-back in the azimuthal angle, with identical p_T (to obey to the momentum conservation). For most of the common jet definitions, this is a sufficient condition to avoid any merging of the two partons in one final jet. Each final state parton is identified as a final jets, with no differences due to the jet definition in use.

A final state with only two partons is clearly a rough approximation, because the multiplicity of particles in the measured events can be really high (much bigger than 2 particles). The discrepancy in the multiplicity of the final state objects is reduced by the use of a jet algorithm. Nevertheless, the measured jets are complex QCD observables, and different jet definitions produce jets with different properties in the measurement, reinforcing the need of a more accurate description of the final state system in the theoretical prediction.

Another weak point in the LO prediction is the intrinsic uncertainty in the perturbative expansion. This is a general aspect for the perturbative calculations, and will be discussed in the next section.

2.2. Higher orders

The hope to improve the accuracy of the perturbative predictions pushes to provide the calculation of the higher order terms in the expansion. The next term, after the leading, is the Next to Leading Order (NLO). To estimate the NLO cross section, two different and divergent terms must be calculated: the real emission of an extra parton, and the virtual contribution. These two contributions diverge in some configurations (infrared and collinear limits), but if they are summed together in the proper way, they counterbalance. This requires an infra-red and collinear safe observable. For the inclusive jet cross section, the jet definition plays this role. This proceeding does not report a description of the problems caused by the selection of an infrared and/or collinear unsafe jet algorithm (it can be found in several references, i.e. [7]).

By using an infra-red and collinear safe jet definition, the NLO prediction for the inclusive jet cross section can be calculated with NLOJET++ [4] (Figure 1). In this figure, proton-proton collisions at $\sqrt{s}$ =7 TeV are simulated by using the CTEQ6.5 PDF [5], and the anti-k_T with R=0.6 to reconstruct the jets [6]. The renormalization and the factorization scales are equal to the leading jet p_T.

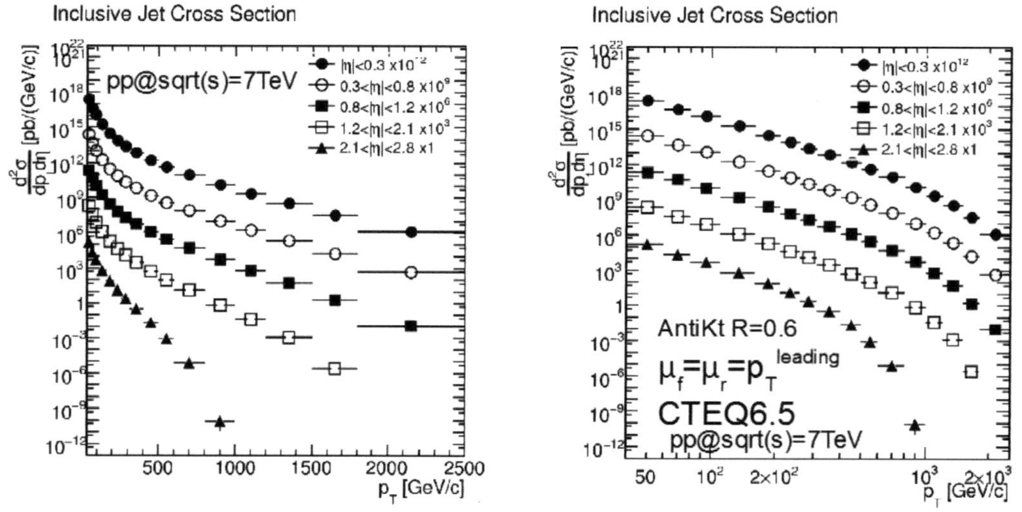

Figure 1. Inclusive jet cross section estimated at the next to leading order with the anti-k_T jet algorithm with R=0.6. Left: vertical logarithmic scale. Right: double logarithmic scale.

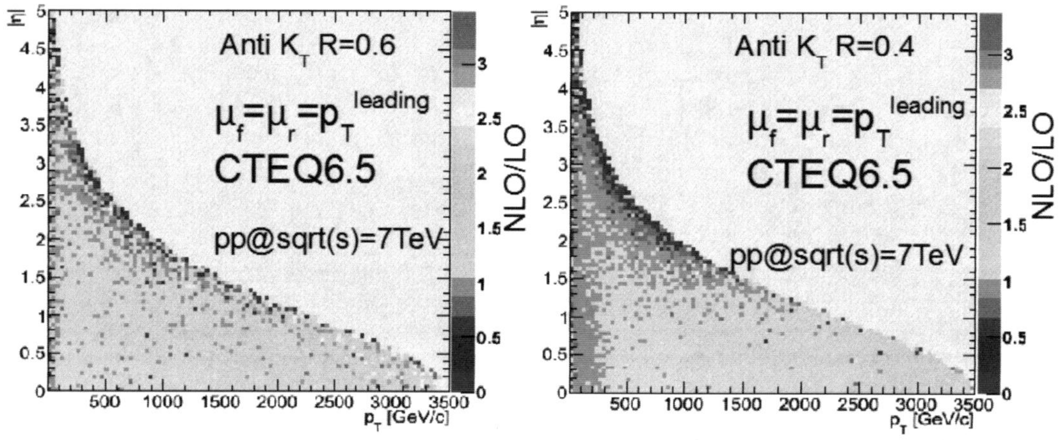

Figure 2. Ratio NLO/LO for different point of the final state phase space for the inclusive jet cross section, for anti-k_T jets with R=0.4 and R=0.6.

2.3. Comparing Leading Order and Next to leading Order

The use of the second term in the expansion could change the prediction in some special point of the phase space. To check the difference between the leading order and the next to leading order prediction, the ratio of the two cross sections in bins of rapidity and p_T is shown in Figure 2.

In this plot, two different radii for the anti-k_T jet algorithm are used: R=0.6 on the left, and R=0.4 on the right. To reduce the effect of the PDF, the CTEQ 6.5 is used for both the LO and for the NLO predictions. In most of the phase space, the ratio is close to 1, confirming that the contribution of the second term is small, for the selected scales. In some regions, close to the end of the phase space, the difference is bigger. The radius R=0.4 seems to be slightly more affected by the contribution of the second order.

2.4. Stability of the fixed order predictions

The LO (NLO) prediction is the perturbative expansion truncated to the first (second) non trivial order. The accuracy of a truncated expansion (in this case at the first or second order) is good if the contributions from the higher orders are small.

A rough estimate of the goodness of the truncated series can be obtained from the stability of the expansion, when shifting the renormalization (μ_r) and factorization (μ_f) scale. These scales are usually representative of the energy scale of the process. The final prediction should not depend on the selection of these scales (since there is no correspondence in the measured observable), but a spurious dependence remains in the truncated perturbative expansion. One can reasonably hope to see a smaller dependence and a better stability in including further orders in the expansion.

No strong and fundamental argument fixes these scales to a precise and defined energy for a certain observable, and different definitions are possible. For the inclusive jet cross section, the scales can be of the order of the p_T of the leading jet and a possible variation can be $0.5 \times p_{T,max} < \mu_r, \mu_f < 2 \times p_{T,max}$.

To check the stability of the inclusive jet cross section, its dependence as a function of the renormalization and factorization scale variations is shown in Figure 3 on the left. This figure shows the dependence of the inclusive jet cross section $\int dp_T d\sigma/dp_T$ integrated in the p_T interval between 170 GeV and 210 GeV and in the pseudo-rapidity range $|\eta| < 5.0$. The calculation, performed with NLOJET++, is done for proton-proton collisions at $\sqrt{s} = 14$ TeV[1]. The PDFs for the NLO calculation is the CTEQ6.1 and for the LO the CTEQ6.LL.

The scales for the calculation is proportional to the p_T of the leading jet, and three different schemes of variation are used for the LO and the NLO predictions:

$$\begin{aligned} \mu_f = \mu_r = \alpha \times p_{T,max} \\ \mu_r = \alpha \times p_{T,max},\ \mu_f = p_{T,max} \\ \mu_f = \alpha \times p_{T,max},\ \mu_r = p_{T,max} \end{aligned} \tag{1}$$

in which both the scales are simultaneously or independently shifted.

The shift of the scales for the LO prediction produces a drastic variation of the prediction, and the introduction of the next order in the expansion improves the stability of the result.

The envelops of the three schemes of variation reported above, limiting α in the range $[0.5, 2]$ are shown as a function of p_T in Figure 3,right. In this plot, the variations are normalized to the NLO prediction with $\mu_f = \mu_r = p_{T,max}$. The width of the bands is an estimate of the stability of the prediction, and it shows the reduction of uncertainty from approximately 30-40% at the LO to a 5-10% at the NLO.

As expected, the corrections due to the second leading order in the perturbative expansion improve the stability of the prediction.

In addition, the real emission of an extra final parton in the NLO is a first step to overcome the simplicity of the final state discussed above for the LO. The jets in the NLO calculation of the inclusive jet cross section have an internal structure, formed by one or two partons.

3. Some approximations: the parton showers

The NLO calculation provides an accurate prediction as long as the observables are infra-red and collinear safe (i.e. jets), but it is not describing of the production of particles (i.e. pions or kaons) in the jets.

[1] The major features of the LO and of the NLO stability of the cross sections due to renormalization and factorization scale variation marginally depend on the center of mass energy.

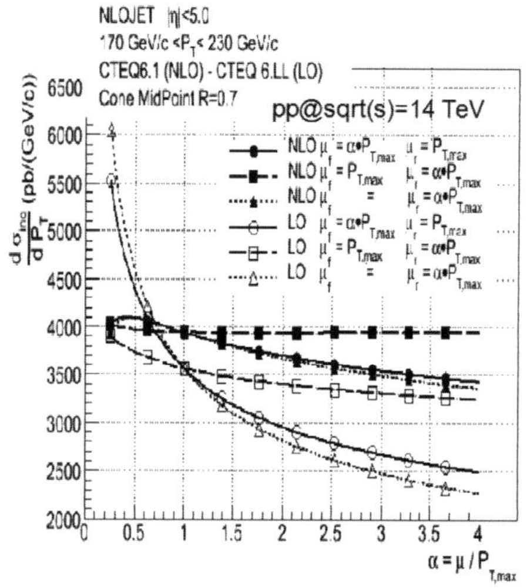

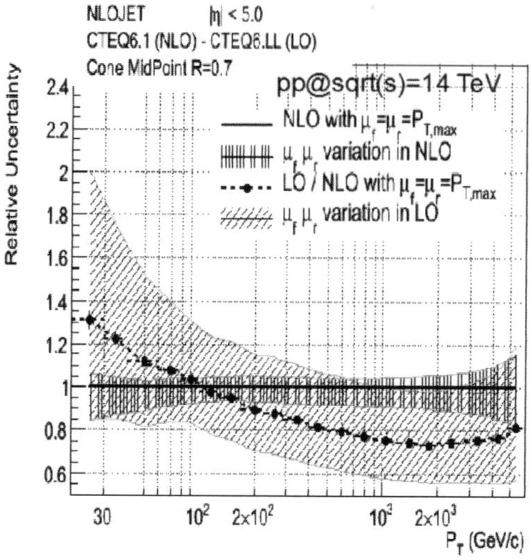

Figure 3. Dependence of the LO and NLO inclusive jet cross section on the renormalization and factorization scale. Left: variation of the cross section in the range in p_T [170 GeV, 210 GeV] for different schemes of scale variation. Right: stability bands for the scale variation as a function of p_T for the inclusive jet cross section, normalized to the NLO prediction with $\mu_f = \mu_r = p_{T,max}$.

The divergences in the real emission of an extra parton can be interpreted as the footprint of the high probability to emit gluons in the collinear and soft regions. With this idea, one can perform an approximated sum over all the perturbative orders to calculate the probability to emit collinear or soft gluons. Whereas the real emission of one extra gluon have a bare infinity if one took the collinear and soft limit, in this approximated calculation, the emission probability is simply bounded to be between 0 and 1.

Several code were proposed and used to simulate this approximated emission of partons [8, 9, 10, 11, 12, 13]. These code recursively simulate the emission of multiple partons, producing a multi parton final state. This gives a parton shower event. To get a prediction for the inclusive jet cross section, the parton shower can be merged/matched with the LO or the NLO matrix elements. Different Monte Carlo generators can produce parton shower events for the jet production [8, 9, 10, 11, 13] starting from the LO matrix element.

The results for the inclusive jet cross section simulated with Pythia 6.421 [8] at the level of the parton shower are shown in Figure 4,left. In this plot the cross section is calculated in different regions of rapidity for the anti-k_T R=0.6 jet algorithm.

Since the parton shower is not changing the normalization of the cross section, the accuracy of the calculation is still at the LO. But the simulation of the parton shower is changing the structure of the jets and the final shape of the spectra. In particular the parton shower changes the integrated energy in the catchment area of the jet. The jets produced by the anti-k_T jet algorithm with R=0.6 integrate more energy in their catchment area with respect to a smaller radius (R=0.4). The impact of the parton shower for these two radii is shown in Figure 4,right. In this plot, the predictions of the inclusive jet cross section for the anti-k_T R=0.6 and R=0.4, are divided by the pure LO prediction. Since the pure LO prediction does not depend on the parameter R, the difference between R=0.4 and R=0.6 is due to the parton shower. The anti-k_T with R=0.6 is producing an higher cross section, simply for the effect of the radiation of gluons

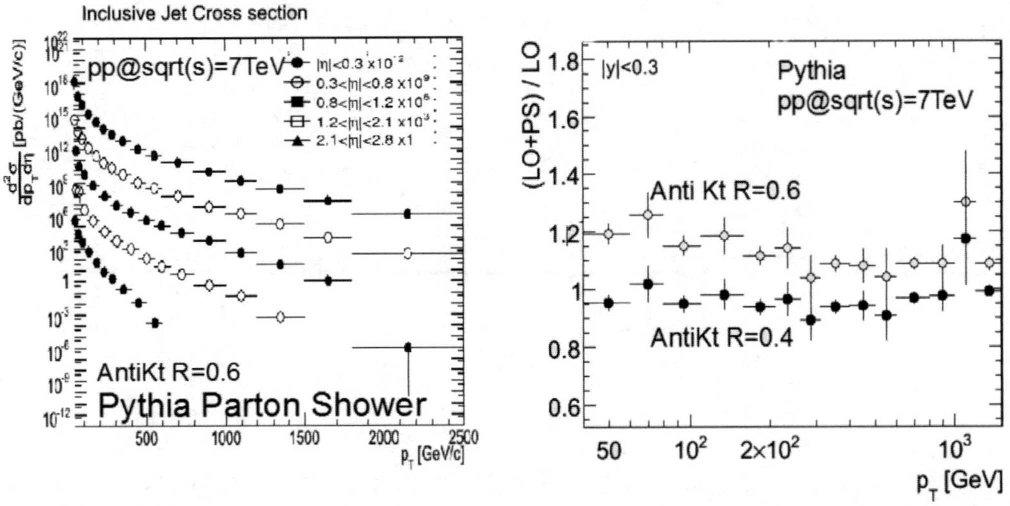

Figure 4. Left: Inclusive jet cross section estimated by Pythia at the end of the parton shower for the anti-k_T jet algorithm with R=0.6. Right: Comparison of the inclusive jet cross section at the end of the parton shower simulation for anti-k_T with R=0.4 and R=0.6. The simulation is performed in the central rapidity region $|y|$ <0.3 with the Pythia Monte Carlo simulation. The two estimate are normalize to the LO parton level prediction obtained byt Pythia.

in the parton shower.

3.1. Non perturbative effects

Even if the transition between the parton level and the hadron level (hadronization) is not completely understood, different models have been developed. The Monte Carlo generators can produce final states which are formed by hadrons. In addition to the hadronization, the Monte Carlo generators have models which take into account the effects due to the proton remnants, which are not directly taking part to the hard scattering. The effects of these "spectators" (which are color connected with the rest of the event, and could eventually produce a second hard scattering) are named underlying event effects. To reproduce as well as possible the properties of the particle production at hadron colliders, these models are tuned to some measured observables.

The non perturbative effects have an impact on the final cross section. To separate these effects from the parton shower effects, the cross section at hadron level σ_{Had} is compared to the one obtained at the end of the parton shower simulation σ_{PS}. The ratio:

$$C_{NP} = \sigma_{Had}/\sigma_{PS} \tag{2}$$

is called non perturbative correction.

Since different models of parton showers, of hadronization and of underlying event exists, different corrections can be obtained. Figure 5 shows the non perturbative corrections for the inclusive jet cross section at $\sqrt{s} = 7$ TeV for the anti-k_T jet algorithm with R=0.4 and R=0.6 in the region in rapidity $|y| < 0.3$. Different tunes and Monte Carlo generators (Pythia [8] and Herwig [10]) are used to derive these results[2].

[2] The results are obtained by running the Monte Carlo generators with the Agile interface in the Rivet framework.

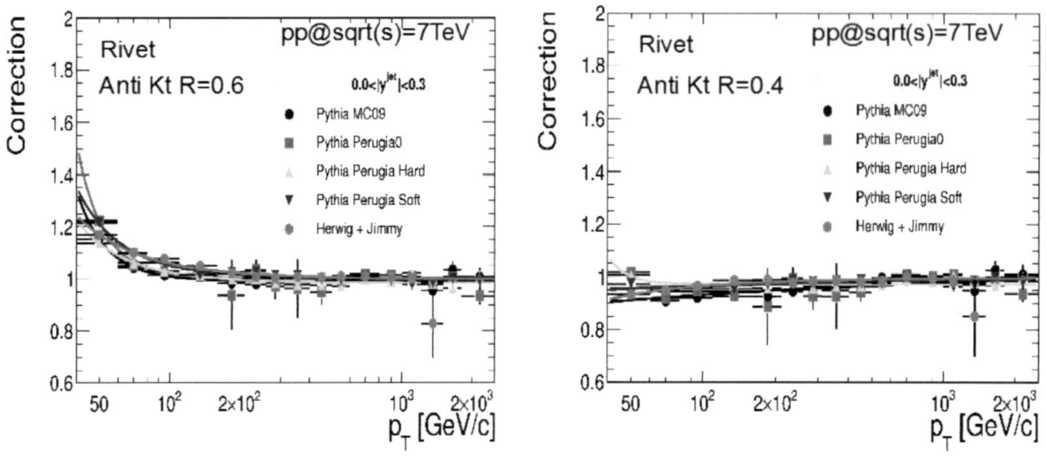

Figure 5. Non perturbative corrections for the inclusive jet cross section for different Monte Carlo generators and tunes. Left: anti-k_T jet algorithm with R=0.6. Right: anti-k_T jet algorithm with R=0.4.

The corrections are really close to 1 for p_T above 200 GeV, which means that in this regime the effect of the hadronization and underlying event is small. The effect of the hadronization and underlying event depends on the radius R of the anti-k_T jets.

4. A solid prediction?

Even if the NLO prediction is not producing a complete list of particles in the final state, and the jet substructure is produced by the emission of only one extra parton, the calculation is usually compared with the measurement. The non perturbative correction, shown in Figure 5 can be used to overcome the lack in the description of the hadronization and the underlying event. These corrections are used as a bin-by-by factors to be multiply to the NLO cross section.

These factor are useful for the comparison of the measured cross section with the theoretical predictions, and to perform QCD fits (i.e. the PDF fits). To avoid any source of ambiguity, an agreement on how to define the factors, on how to estimate their uncertainties and on how to use them would be really useful. For example, in the region where the correction are big, (i.e. 1.5) the perturbative calculation is not accurate, and the contribution from the non perturbative physics is important. Can the NLO prediction, corrected with the non perturbative factor, be trusted as an accurate estimate of the cross section in these regions? Which is the proper way to estimate the theoretical uncertainties?

4.1. A coherent prediction

To overcome the weak points of using a non perturbative correction, one needs to perform a NLO calculation, merged (or matched) with a parton shower which can be interfaced with the hadronization and the underlaying event models.

In the last years several important progresses were done in this direction. The two main formalisms to perform this coherent predictions are MC@NLO [14] and Powheg [15]. Only recently the dijet production was introduced in the Powheg formalism, and it can be used as a natural benchmark to test the method and to check the predictions.

Other processes are calculated with these formalisms, but only few of them have two real final partons in the leading order contribution. The checks on the jet production is important in the optic of moving to other processes with two or more jets in association with other objects in the final state.

5. Conclusion

In this proceeding, several tools to estimate the inclusive jet cross section at hadron colliders have been discussed. The goal of the discussion is not a precise report of all the different aspect of the available calculations, but the review of some aspects of the different approaches.

The advent of the NLO predictions, which give a more stable prediction with respect to the LO, merged with the parton shower approach, which approximately describes the emission of extra partons, is one of the main improvement of the last years. These simulations are easily interfaced with the hadronization models and underlying event models, which describe the non perturbative part of the proton-proton scattering.

This new approach to obtain a theoretical prediction for QCD processes is really promising, overcoming the weak points of the pure NLO prediction, such as the lack of description of the final state, and the necessity to derive the non perturbative corrections.

Acknowledgments

The author would like to thank the organizer of the conference for the precious opportunity to discuss these aspects. The author would like to thank C. Roda, J. Butterworth, P. Loch, P. Starovoitov, T. Carli, G. Salam, K. Perez, Z. Nagy, M. Seymour, P. Nason, C. Oleari, E. Re, P. Skands T. Sjöstrand and R. Torre for their help and for the useful discussions.

References

[1] ATLAS Collaboration, Measurement of inclusive jet and dijet cross sections in proton-proton collisions at 7 TeV centre-of-mass energy with the ATLAS detector, Eur. Phys. J. C71 (2011) 1512, arXiv:1009.5908 [hep-ex].

[2] ATLAS Collaboration, Measurement of inclusive jet and dijet cross sections in proton-proton collision data at 7 TeV centre-of-mass energy using the ATLAS detector, ATLAS-CONF-2011-047

[3] CMS Collaboration, Measurement of the Inclusive Jet Cross Sectionin pp Collisions at $\sqrt{s}=$ 7 TeV, arXiv:1106.0208v1 [hep-ex].

[4] Z. Nagy, Next-to-leading order calculation of three jet observables in hadron hadron collision, Phys. Rev. D68 (2003) 094002, arXiv:hep-ph/0307268.

[5] W. K. Tung, H. L. Lai, A. Belyaev, J. Pumplin, D. Stump and C. P. Yuan, JHEP **0702** (2007) 053 [arXiv:hep-ph/0611254].

[6] M. Cacciari, G. Salam, and G. Soyez, The anti-kt jet clustering algorithm, JHEP 0804 (2008) 063, arXiv:arXiv:0802.1189.

[7] C. Buttar et al., Standard Model Handles and Candles Working Group: Tools and Jets Summary Report, arXiv:arXiv:0803.0678 [hep-ph].

[8] T. Sjostrand, S. Mrenna, and P. Skands, PYTHIA 6.4 physics and manual, JHEP 05 (2006) 026, hep-ph/0603175.

[9] T. Sjostrand, S. Mrenna and P. Skands, Comput. Phys. Commun. 178 (2008) 852 [arXiv:0710.3820 [hep-ph]]; `http://home.thep.lu.se/~torbjorn/Pythia.html`.

[10] G. Corcella et al., JHEP 0101 (2001) 010 [arXiv:hep-ph/0011363]; `http://hepwww.rl.ac.uk/theory/seymour/herwig/`.

[11] M. Bahr et al., Eur. Phys. J. C58 (2008) 639 [arXiv:0803.0883 [hep-ph]]; `http://projects.hepforge.org/herwig/`.

[12] L. Lonnblad, Comput. Phys. Commun. 71 (1992) 15.

[13] T. Gleisberg, S. Hoche, F. Krauss, M. Schonherr, S. Schumann, F. Siegert and J.Winter, JHEP 0902 (2009) 007 [arXiv:0811.4622 [hep-ph]]; `http://projects.hepforge.org/sherpa/`.

[14] S. Frixione, F. Stoeckli, P. Torrielli, B. R. Webber, and C. D. White, The MC@NLO 4.0 Event Generator, arXiv:1010.0819 [hep-ph].

[15] S. Alioli, P. Nason, C. Oleari, and E. Re, A general framework for implementing NLO calculations in shower Monte Carlo programs: the POWHEG BOX, arXiv:1002.2581 [hep-ph].

[16] S. Alioli, K. Hamilton, P. Nason, C. Oleari, and E. Re, Jet pair production in POWHEG, arXiv:1012.3380 [hep-ph].

Experimental bounds on large extra dimensions from di-jet event production in hadron collisions

Pier Paolo Giardino[a] **and Paolo Lodone**[b1]
[a] Università di Pisa and INFN, Largo Bruno Pontecorvo 3, 56127 Pisa, Italy
[b] Scuola Normale Superiore and INFN, Piazza dei Cavalieri 7, 56126 Pisa, Italy

E-mail: pierpaolo.giardino@pi.infn.it , p.lodone@sns.it

Abstract. We discuss the new dominant bounds that can be derived on the coefficient of the effective operator generated by tree-level graviton exchange in large extra dimensions from pp $\to$ jj data at LHC: $M_T > 2.1$ TeV (ATLAS after 3.1/pb of integrated luminosity), $M_T > 3.3$ TeV (ATLAS after 36/pb of integrated luminosity), $M_T > 3.4$ TeV (CMS after 36/pb). We clarify the role of on-shell graviton exchange and compare the full graviton amplitude to data, setting bounds on the fundamental quantum-gravity scale.

1. Introduction

With the start of the LHC program, experiments are already testing directly some of the theoretical ideas about new physics at the electroweak scale. In one popular scenario Standard Model fields are confined on a 3-dimensional brane, while gravity propagates in the full D-dimensional space, with δ flat and compactified extra spatial dimensions ($D = 4 + \delta$) [1]. This scenario allows for quantum gravity at the weak scale and could therefore be a solution to the Higgs mass hierarchy problem. Even without knowledge of the exact model for quantum gravity at the weak scale, we can make some definite predictions for collider experiments which can provide valid descriptions in certain kinematical domains. We can identify five different kinds of LHC signals which allow for a theoretical interpretation in terms of D-dimensional gravity (for details we refer to [2]).

(i) *Missing p_T from emission of massive gravitons* constituting the Kaluza-Klein tower.

(ii) *Tree-level exchange of gravitons* generating the effective dimension-8 operator $\mathcal{T}$ [3, 4, 5]

$$\mathscr{L}_{\text{int}} = c_{\mathcal{T}} \times \mathcal{T} = \frac{8}{M_T^4} \times \frac{1}{2}\left(T_{\mu\nu}T^{\mu\nu} - \frac{T^{\mu}_{\mu}T^{\nu}_{\nu}}{\delta+2}\right), \tag{1}$$

where $T_{\mu\nu}$ is the SM energy-momentum tensor. In most cases the parameter M_T cannot be computed without knowledge of the underlying quantum-gravity theory.

(iii) *Virtual graviton exchanges at one-loop level* induce dimension-6 effective operators, as opposed to the dimension-8 $\mathcal{T}$ operator [6]. For pure graviton virtual intermediate states, a unique dimension-6 operator is generated

$$\mathscr{L} = c_\Upsilon \times \Upsilon, \quad \Upsilon = \frac{1}{2}\Big(\sum_f \bar{f}\gamma_\mu\gamma_5 f\Big)\Big(\sum_f \bar{f}\gamma^\mu\gamma_5 f\Big), \tag{2}$$

[1] Speaker.

where f is any SM quark or lepton. As in the case of tree-level graviton exchange, the coefficient c_Υ can be related to the fundamental parameters M_D and δ only by specifying a cutoff procedure.

(iv) *Dijet events at large invariant mass and large rapidity separation.* In this kinematic regime, gravitational scattering can be reliably computed in the eikonal approximation [7].

(v) *Black holes.* Black-hole formation and decay is expected to occur in the transplanckian region when the impact parameter becomes smaller than the corresponding Schwarzschild radius [8].

Experiment	Process	+	−
LEP [9]	$e^+e^- \to \gamma\gamma$	0.93 TeV	1.01 TeV
LEP [10]	$e^+e^- \to e^+e^-$	1.18 TeV	1.17 TeV
H1 [11]	e^+p and e^-p	0.74 TeV	0.71 TeV
ZEUS [12]	e^+p and e^-p	0.72 TeV	0.73 TeV
CDF [13]	$p\bar{p} \to e^+e^-, \gamma\gamma$	0.99 TeV	0.96 TeV
DØ [13]	$p\bar{p} \to e^+e^-, \gamma\gamma$	1.28 TeV	1.14 TeV
DØ [14]	$p\bar{p} \to jj$	1.48 TeV	1.48 TeV
CMS at 7 TeV with 40/pb [15]	$pp \to \mu^-\mu^+$	1.6 TeV	1.6 TeV
CMS at 7 TeV with 36/pb [16]	$pp \to \gamma\gamma$	1.74 TeV	1.71 TeV
ATLAS at 7 TeV with 3.1/pb	$pp \to jj$	2.2 TeV	2.1 TeV
ATLAS at 7 TeV with 36/pb	$pp \to jj$	4.2 TeV	3.2 TeV
CMS at 7 TeV with 36/pb	$pp \to jj$	4.2 TeV	3.4 TeV

Table 1. Tree-level graviton exchange: *95% CL limits on the coefficient M_T (known as Hewett normalization [5]) of the dimension-8 operator $\mathcal{T}$ of eq.* (1) *for positive and negative interference. The last three limits are derived in [2].*

Experiment	Process	+	−
LEP combined [17]	$e^+e^- \to e^+e^-$	11.3	11.5
LEP combined [17]	$e^+e^- \to \mu^+\mu^-$	16.4	12.7
LEP combined [17]	$e^+e^- \to \ell^+\ell^-$	17.2	15.1
LEP combined [17]	$e^+e^- \to b\bar{b}$	15.3	11.5
H1 [11]	e^+p and e^-p	2.5	3.9
ZEUS [12]	e^+p and e^-p	4.6	5.3
DØ [18]	$p\bar{p} \to e^+e^-$	4.7	5.5
CDF [18]	$p\bar{p} \to \ell^+\ell^-$	4.5	5.6
CCFR [19]	νN scattering	3.7	5.9
DØ [18]	$p\bar{p} \to jj$	3.2	3.1
ATLAS at 7 TeV with 3.1/pb	$pp \to jj$	5.3	4.2
CMS at 7 TeV with 36/pb	$pp \to jj$	11	8.1
combined		22.4	15.7

Table 2. Loop-level graviton exchange: *95% CL limits on the coefficient $|c_\Upsilon/4\pi|^{-1/2}$ (in* TeV*) of the dimension-6 operator Υ of eq.* (2) *for positive and negative values of c_Υ.*

In its first stage with low statistics, LHC is particularly sensitive to the operator in eq. (1), because its high dimensionality means that the high energy of the LHC collisions is the key factor.

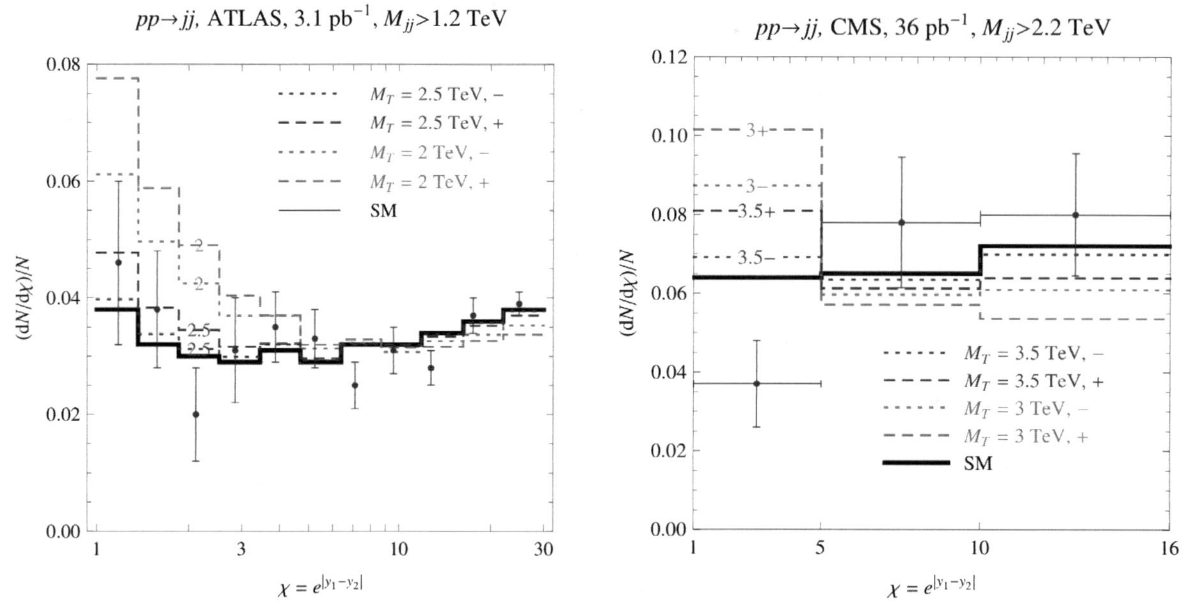

Figure 1. *Left (right): $pp \to jj$ angular distribution at ATLAS with $M_{jj} > 1.2\,\mathrm{TeV}$ (at CMS with $M_{jj} > 2.2\,\mathrm{TeV}$) binned as a function of the angular distance χ. The experimental data (crosses) are compared to the SM prediction (black histogram) and to the expectation including virtual graviton effects at tree level.*

2. Fit to the graviton-exchange effective operator

We compare the first LHC data to the new physics described by eq.s (1) and (2). The tree-level exchange of virtual gravitons described by the Lagrangian of eq. (1) mediates the processes: $pp \to \ell^+\ell^-$, $pp \to \gamma\gamma$, $pp \to jj$. The main point is that the $pp \to \ell^+\ell^-$ and $pp \to \gamma\gamma$ cross sections are significantly lower than the $pp \to jj$ cross section. Indeed requiring final states with invariant mass greater than 1 TeV, jets, leptons and photons with $\eta < 2.5$, and additionally requiring $|\eta_1 - \eta_2| < 1.2$ for the jets, we find:

$$\sigma = \left(\frac{2\,\mathrm{TeV}}{M_T}\right)^8 \times \left\{ \begin{array}{ll} 12.5\,\mathrm{pb} & \text{for } pp \to jj \\ 10.4\,\mathrm{fb} & \text{for } pp \to \mu^+\mu^- \\ 21.3\,\mathrm{fb} & \text{for } pp \to \gamma\gamma \end{array} \right. . \tag{3}$$

This large difference in cross sections is due partly to trivial flavor and color factors, and partly to the fact that the processes are mediated by the operator of dimension 8 in eq. (1). In particular $pp \to jj$ benefits from the high energy of the initial partons uu in the t-channel process.

ATLAS [20] and CMS [21, 22] have searched for the effect of contact interactions in the angular distribution of dijet events, in the variable $\chi \equiv \exp|y_1 - y_2|$ where $y_{1,2}$ are the two jet rapidities. Due to the dominance of Coulomb-like scattering in the SM, these distributions are expected to be almost flat in the case of QCD, while contact interactions give a deviation from the flat distribution. Data for the χ distribution from ATLAS are reported in fig. 1a together with the SM expectation at next-to-leading order [20]. We compare data with the theoretical expectation and we compute the 95% CL bound on the coefficientof the $\mathcal{T}$ operator. We find the bound $M_T > 2.1\,\mathrm{TeV}$ reported in table 1, which significantly exceeds all previous bounds. CMS $pp \to jj$ data after 36 pb^{-1} have been presented in [22] and are here plotted in fig. 1b. We can reliably estimate the resulting bound, $M_T > 3.4\,\mathrm{TeV}$, as reported in table 1. From ATLAS $pp \to jj$ data after 36 pb^{-1} [23] we estimate $M_T > 3.2\,\text{TeV}$.

Finally, we computed the bound on the dimension-6 operator Υ of eq. (2) generated by graviton exchange at loop level. The result is shown in table 2 together the other existing bounds. With the published data we find a bound from dijets at LHC that is comparable to the bound from Tevatron and strongly subdominant with respect to the bound from LEP.

3. Fit to the full graviton-exchange amplitude

In view of its experimental significance, in [2] the theory behind eq. (1) is recosidered. In full generality, tree-level graviton-exchange leads to a scattering amplitude of the form

$$\mathcal{A} = \mathcal{S}(s)\left(T_{\mu\nu}T^{\mu\nu} - \frac{T^{\mu}_{\mu}T^{\nu}_{\nu}}{\delta+2}\right), \qquad (4)$$

where the function $\mathcal{S}$ is obtained by summing over all the Kaluza-Klein (KK) tower of gravitons. If the typical energy resolution of the experiment is broader than the mass separation between two KK states, then the sum can be approximated with an integral over the extra-dimensional momentum q of the graviton. In the opposite case of well separated narrow resonances, it is more precise to square the amplitude before integrating. This guess is confirmed by comparison with the exact result for the sum in the cases in which it can be explicitly performed. This gives an enhancement which however corresponds to on shell graviton emission, and thus must be retained only when those decay promptly in the detector. On the other hand it actually corresponds to missing energy and not dijet signal if the emitted gravitons decay far away. All this is clarified in [2], to which we refer for more details.

The integral in $\mathcal{S}$ is UV divergent for $\delta > 1$ extra dimensions, so we regularize them by including only KK excitations with mass $m = |q|$ below an arbitrary cut-off Λ, which parametrizes the onset of the unknown quantum-gravity physics. A small (large) ratio Λ/M_D effectively means that quantum gravity is weakly (strongly) coupled [6]. In view of the high dimensionality of the operator, the dominant LHC bound comes from the highest energy events, and it is appropriate to retain the full amplitude, including the dependence on the cut-off Λ. Formulæ for the cross sections from tree-level graviton effects in any number of extra dimensions can be found in the appendix of ref. [24]. We implement them in PYTHIA8 [25] and compare the data with the full graviton-exchange amplitude. The results of our fit are shown in fig. 2.

4. Conclusion

We found that the very first LHC data about $pp \to jj$ improve significantly previous bounds on the coefficient of the effective dimension-8 operator $\mathcal{T}$ generated by virtual graviton exchange, and predicted by theories with extra dimensions.

We also went beyond the effective-operator approximation and computed the full amplitude generated by tree level graviton exchange in terms of a cut-off parameter Λ, which is the maximal KK graviton mass. We clarified that the enhanced effect of lighter gravitons that can be produced on-shell must be included only when such gravitons decay within the detector. Fig. 2 shows the resulting LHC bounds in the $(M_D, \Lambda/M_D)$ plane.

Acknowledgments

The work of PL is supported in part by the EU ITN "Unification in the LHC Era", contract PITN-GA-2009-237920 (UNILHC), by the MIUR under contract 2008XM9HLM.

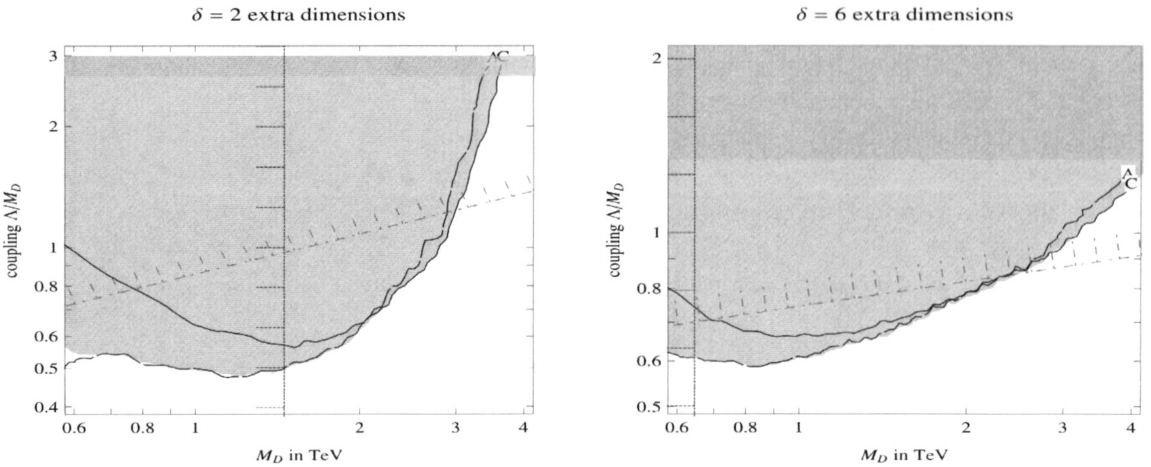

Figure 2. *The shaded area is the bound from virtual graviton exchange at CMS (continuous line denoted as 'C', data after* 36/pb*), ATLAS (long-dashed line denoted as 'A', data after* 36/pb*). Vertical blue line: bound from graviton emission (as summarized in table 1 of [6]). Red line: Naive Dimensional Analysis estimate of LEP bound from loop graviton exchange. Upper shading: NDA estimate of the non-perturbative region.*

References

[1] N. Arkani-Hamed, S. Dimopoulos and G. R. Dvali, Phys. Lett. B 429 (1998) 263 [hep-ph/9803315].

[2] R. Franceschini, G. F. Giudice, P. P. Giardino, P. Lodone and A. Strumia [hep-ph/1101.4919].

[3] G. F. Giudice, R. Rattazzi and J. D. Wells, Nucl. Phys. B 544, 3 (1999) [hep-ph/9811291].

[4] T. Han, J.D. Lykken and R. Zhang, *Phys. Rev.* D59 (1999) 105006 [arXiv:hep-ph/9811350].

[5] J.L. Hewett, *Phys. Rev. Lett.* 82 (1999) 4765 [arXiv:hep-ph/9811356].

[6] G. F. Giudice and A. Strumia, Nucl. Phys. B 663 (2003) 377 [hep-ph/0301232].

[7] G. F. Giudice, R. Rattazzi, J. D. Wells, Nucl. Phys. B630 (2002) 293-325 [hep-ph/0112161].

[8] S. B. Giddings, S. D. Thomas, Phys. Rev. D65, 056010 (2002) [hep-ph/0106219]. S. Dimopoulos, G. L. Landsberg, Phys. Rev. Lett. 87, 161602 (2001). [hep-ph/0106295]. See also T. Banks, W. Fischler, hep-th/9906038.

[9] The LEP-II Diphoton Working Group, LEP2FF/02-02.

[10] ALEPH 2001-019 CONF 2001-016; DELPHI 2001-094 CONF 522; L3 Note 2759; OPAL Physics Note PN471.

[11] H1 Collaboration, ICHEP02, abstract 979.

[12] ZEUS Collaboration, EPS 2001, abstract 602.

[13] G. Landsberg, for the D0 and CDF collaborations, arXiv:hep-ex/0412028.

[14] D0 collaboration, Phys. Rev. Lett. 103 (2009) 191803 [arXiv:0906.4819].

[15] CMS collaboration, `http://cdsweb.cern.ch/record/1335097/files/EXO-10-020-pas.pdf`.

[16] CMS collaboration, arXiv:1103.4279. See also: CMS-PAS-EXO-09-004, "Search for Large Extra Dimensions in the Diphoton Final State".

[17] LEP Working Group LEP2FF/02-03.

[18] J.A. Green, for the CDF and D0 Collaborations, arXiv:hep-ex/0004035.

[19] K.S. McFarland *et al.* [NuTeV Collaboration], arXiv:hep-ex/9806013.

[20] ATLAS collaboration, *Phys. Lett.* B694 (2011) 327 [arXiv:1009.5069].

[21] CMS collaboration, *Phys. Rev. Lett.* 105 (2010) 262001 [arXiv:1010.0203].

[22] CMS collaboration, arXiv:1102.2020. See also the talk by G. Landsberg, "Quest for New Physics with the First LHC data at CMS", 24/1/2011.

[23] ATLAS collaboration, arXiv:1103.3864.

[24] G. F. Giudice, T. Plehn, A. Strumia, Nucl. Phys. B706 (2005) 455 [arXiv:hep-ph/0408320].

[25] PYTHIA: T. Sjostrand, S. Mrenna, P. Skands, Comput. Phys. Com. 178, 852 (2008) [arXiv:0710.3820].